Pyramids, Plato, and Planets

A Brief History of Math and Science

Book 2
Egypt,
and Babylonia

Fred Benham

D1176906

Pyramids, Plato, and Planets

A Brief History of Math and Science

Book 2
Egypt
and Babylonia

Fred Benham

Illustrations by the Author

Illustrations have been formatted
to be legible on handheld displays.

Preface

Egypt and Babylonia is book two in a series of nine books covering the history of math and science. The series is intended to bridge the gap between the too general and the too detailed. It is intended for the general audience but also it contains sufficient information so that it can be used as a top level textbook. Numerous illustrations are presented to help guide the reader through each topic. The names "Babylonia" and "Mesopotamia" are both used in this book to address the geographical area between and surrounding the Tigris and Euphrates Rivers, this area corresponds to modern Iraq, Syria, and eastern Turkey.

In order to accurately present certain ideas[1], some algebraic formulas are presented where appropriate (a short algebra refresher is contained in the appendix), however, the material should be of general interest even if the mathematics is bypassed. Also, since math and science were developed over the centuries, the information is provided within the context of an historical timeline.

This book covers the civilizations of Egypt and Babylonia from 3300 BC to 500 BC. By 3300 BC what we refer to as

---

[1] In some volumes of this series, the phrase "since mathematics is the language of science" has been used, However, as pointed out to me by Dr. Dan Varon, there is a concept presented in the book "Our Mathematical Universe" by MIT Physics Professor Max Tegmark which presents the idea that the universe IS a mathematical structure which we are discovering..

"civilization" had developed with a central government, trade with neighbors, religion, writing, agriculture, animal husbandry, systems of measurement, and rudimentary mathematics. The Egyptian timeline is divided into the Old Kingdom c2700 BC during which the famous Pyramids of Giza were built; the Middle Kingdom c2100 BC during which the first mathematical papyruses were written; and the New Kingdom c1600 BC with notables such as Akhenaten, Nefertiti (Queen to Akhenaten), Tutankhamen (son of Akhenaten), and Moses. Egypt, because of its geographic isolation and religious constraints, remained essentially unchanged for thousands of years. In contrast with Egypt, Babylonia was in the crossroads of trade and human migrations which resulted in a diversity of peoples and governments. Babylonia is popularly remembered for the Epic of Gilgamesh (c2000 BC), the Code of Hammurabi (c1758 BC), the Great Ziggurat of Ur (c2000 BC), the twelve signs of the Zodiac (c700 BC), and the Hanging Gardens of Babylon (c600 BC).

The Series

The nine books in this series on Math and Science are listed here.

Book 1, Big Bang and Beyond (13.7 BYA to 3300 BC)

Book 2, Egypt and Babylonia (3300 BC to 500 BC)

Book 3, Pythagoras and the Philosophers (500-323 BC)

Book 4, Euclid, Archimedes, & Astronomers (323-146 BC)

Book 5, The Conics of Apollonius (c200 BC)

Book 6, Ancient Rome (146 BC to 476 AD)

Book 7, The Middle Ages (476 AD to 1453 AD)

The Early Middle Ages (476-1000 AD)
The High Middle Ages (1000-1300 AD)
The Late Middle Ages (1300-1453 AD)

Book 8, The Renaissance Plus (1453 AD to 1687 AD)

The Renaissance (1453-1517)
The Reformation (1517-1648 AD)
The Age of Reason (1648-1687 AD)

Book 9, Newton and Beyond (1687 AD to today)

The Enlightenment (1687-1800 AD)
The Classical Age (the 1800s)
The Modern Age (1900 to today)

Table of Contents

Book 2
Egypt
and Babylonia

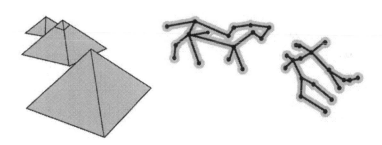

A Prelude to Civilization

The period of time from the origin of the universe to the beginning of civilization is the subject of Book 1 of this series. However, in order to set the stage for the first civilizations, a short overview of this timeframe is presented here.

13.7 billion years ago our universe began in a "Big Bang[1]". The Big Bang theory presumes that the universe (space, time, energy and matter) started out as a marble sized dense cloud of fundamental particles which rapidly expanded and has been expanding ever since. The theory is supported by several observations. First, other galaxies, 200 billion of them, are receding from us and the farther away, the faster they are receding, the implication being that if you go backwards in time, the galaxies converge. Also, the theory successfully predicts that 99.9% of the universe is composed of hydrogen and helium in a ratio twelve Hydrogen atoms to one Helium atom; these ratios are still valid today since only a very small amount of matter has been converted into heavier elements by nuclear fusion within stars; the composition of the Sun is approximately in these ratios. Lastly, the theory predicts that when the universe was very young and very hot it

---

[1] Although the Big Bang theory is a fairly robust and well accepted theory, it only accounts for the "normal" matter and energy we observe; and cosmologists believe the normal matter and energy account for only 4% of the universe and dark matter accounts for 22% and dark energy accounts for 74%.

would have radiated energy much like a hot metal bar turns white hot or red hot depending on the temperature; the theory predicts that this radiant energy should still be measurable, and indeed, the existence of a uniform Cosmic Microwave Background Radiation from space has accurately been measured to be consistent with the theory.

13.2 billion years ago our Milky Way Galaxy was formed. Our Milky Way Galaxy is one of approximately 200 billion other galaxies, each containing an estimated 200 billion stars. The stars in the galaxies come in a variety of sizes. Massive stars burn hotter (and appear whiter); they collapse and explode in a supernova in a relatively short time (e.g. 15 MY) creating heavier elements. The iron in our bodies comes from a supernova. Less massive stars like our sun will continue to burn for another 5 Billion years, after which they will becomes red giants.

4.6 billion years ago our Earth and Solar System were formed. This occurred when a cloud of interstellar gas and particles, located approximately half way between the Milky Way's center and edge, collapsed under the force of gravity. Almost all of the cloud (99.8 percent) collapsed to form our sun, which is mostly hydrogen and helium, the remainder of the cloud formed the planets including of course our Earth.

4.2 billion years ago the building blocks of life appeared. Prior to this time, during the first 400 million years of our Solar System's existence, the planets and moons served as gravitational vacuum cleaners attracting debris that was present after the collapse of the interstellar cloud. As a result the planets and moons were continuously impacted by meteorites. Despite frequent impacts, the molecules and functions necessary for life came into being. How this occurred is the subject of much speculation. One theory, named Panspermia, proposes that the necessary constituents of life can survive space and that the earth was seeded from outer space. Another theory proposes abiogenesis, a process of forming life from inorganic matter. Regardless of the theory, there are serious "chicken and egg" issues for which solutions are not known.

4.0 billion years ago continent formation began. When the Earth was formed the heat associated with gravitational collapse as well as heat due to fission of radioactive elements caused the Earth's interior to be molten, and it remains molten today. The viscous mantle (the Earth's layer between the crust and the core), is in slow but constant motion due to the Earth's rotation. The movement of the mantle pushed the lighter crust, floating on the surface of the mantle, around the globe eventually forming the continents and continental plates. The continental plates currently move at an average speed of about one inch per year, but four billion years ago the mantle was hotter and more fluid so the plates moved faster.

4.0 billion years ago first life appeared. First life was bacterial. Bacteria are single-celled organisms without a nucleus that reproduce by binary fission, they are referred to as prokaryotes. The word prokaryote is from the Greek "pro" meaning before and "karyon" meaning kernel, i.e. before the nucleus. The first bacteria were amazingly complex organisms; they used the same genetic code which is common to all life. The first cell had to: regulate gene expression; synthesize proteins for expressed genes; ingest, digest, and egest; repair and grow the cell membrane; build cell components and divide the cell; cleanup cell debris; and, repair and replicate the DNA. Life and "cells" are synonymous; you can't have one without the other. All life is descended from this first cell. For the next 2.5 billion years, until 1.5 billion years ago, all life on Earth was single-celled.

3.5 billion years ago oxygen photosynthesis evolved. Photosynthesis is a process which takes carbon dioxide, water, and solar energy to form carbohydrates, i.e., CO_2 + H_2O + photons = CH_2O + O_2. Photosynthesis evolved within the blue-green cyanobacteria. The photosynthesis process yields oxygen as a waste product. For a billion years the oxygen from photosynthesis combined with iron and other elements until the need was exhausted, only then did atmospheric oxygen buildup begin.

2.5 billion years ago atmospheric oxygen buildup began. Prior to this time the atmosphere had no oxygen. As the oxygen built up in the atmosphere cellular life finally evolved that could metabolize oxygen which produces much more energy.

1.8 billion years ago cells with a nucleus appeared. Cells with a nucleus are called eukaryotes. The nucleus contains the cell's DNA and having the DNA in one compact region provided an evolutionary advantage by permitting faster chemical reactions. Also, eukaryote cells can metabolize oxygen (using the Krebs cycle) to generate energy. Prior to this, energy was produced using the metabolic pathway called "glycolysis". Generating energy using oxygen is five times more efficient than glycolysis. The evolution of eukaryotes with their higher energy generation was a necessary step to multicellular life.

1.5 billion years ago multicellular life and sexual reproduction evolved. Sexual reproduction requires the cooperation of two different cell types; it is advantageous because it results in varied offspring resulting in a higher survival rate. With sexual reproduction and the ability to create multicellular organisms available, the necessary ingredients were finally available to create a diversity of life forms.

1.5 billion years ago animals, plants, and fungi diverged. Plants, animals, and fungi are all multicellular organisms which have cells with a nucleus, and they all reproduce sexually. The differences are as follows: animals have a means of locomotion whereas plants and fungi do not. Plants have chloroplasts (within which photosynthesis occurs) whereas animals and fungi do not.

1.1 billion years ago the supercontinent Rodinia formed along the equator. At this time North America was rotated about ninety degrees clockwise compared to today.

Australia and Antarctica bordered the North American coast from Oregon to Alaska. South America bordered the North American coast from Florida to Canada. When the supercontinent was formed and South America impacted North America, the Appalachians were formed, higher than the current Himalayas; this is referred to as the Grenville orogeny. Orogeny is the term for mountain building. Rodinia lasted for about 400 million years, until 700 million years ago.

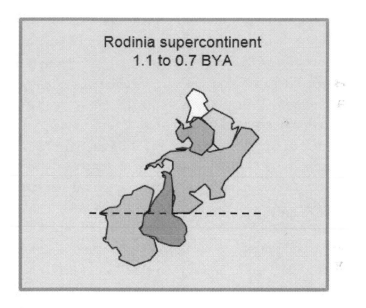

750 million years ago the earth was entirely frozen. This period is referred to as "Snowball Earth". During Snowball Earth the equator was as cold as modern-day Antarctica. The cause of Snowball Earth is not clear but one suggestion is that with the continents all being near the equator the significant rainfall washed calcium rich rock minerals into the ocean. The calcium combined with CO_2

to form calcium carbonate (Limestone) reducing the amount of CO_2 in the atmosphere. The reduced CO_2 reduced the greenhouse effect which lowered the temperature. Because of Snowball Earth, the rapid growth and diversity of multicellular organisms was delayed by 200 million years.

542 million years ago was the "Cambrian Explosion". This was an explosion of life producing the first significant fossil record; this was followed by numerous geological and evolutionary events. The 542 million years are referred to as the Phanerozoic Eon. Phanerozoic is from Greek words meaning visible life. The Phanerozoic is divided into twelve periods, which are, starting with the oldest: Cambrian (54), Ordovician (44), Silurian (28), Devonian (57), Carboniferous (60), Permian (48), Triassic (51), Jurassic (55), Cretaceous (80), Paleogene (42), Neogene (20.4) and Quaternary (2.6). The numbers in parentheses are the durations in millions of years. These periods average about 50 million years in duration, an exception being the most recent period, the Quaternary, at only 2.6 million years. The oldest period is the Cambrian when visible life suddenly appeared, hence the name Cambrian Explosion. The word Pre-Cambrian is often used to refer to the entire four billion year period of time prior to the start of the fossil record. At the beginning of the Cambrian the atmospheric oxygen was about 10 percent, half of the current value; this would continue to build up to the present day value of 20% over the next 100 million years.

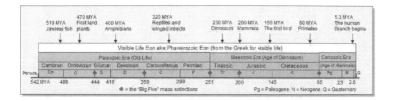

500 million years ago the five senses evolved in fish. From the fish we have inherited the senses of vision, hearing, touch, smell, and taste. It is interesting to note that a human embryo has gill slits and looks much like a fish embryo.

400 million years ago amphibians evolved. This evolutionary step required the development of lungs and four legs. The lungs are speculated to have evolved from air sacs used for buoyancy and the legs evolved from fins which were used to "walk" along the bottom in shallow pools.

300 million years ago the supercontinent Pangaea formed. All the major continents merged into the super continent Pangaea (after 150 million years the continents separated). When Africa and North America collided the Appalachians were pushed up again; they eroded to small hills within 200 million years but were uplifted during the late Cenozoic (about 10 million years ago). The cause of the Cenozoic uplift is the subject of speculation, but isostatic rebound is one possibility. This is when the crust bobs up and down, over millions of years, until equilibrium with the viscous mantle is reached.

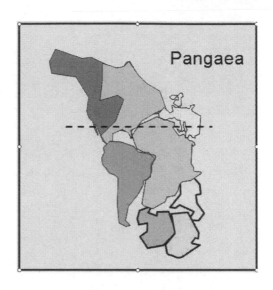

251 million years ago was the Permian-Triassic extinction event, the worst extinction event in Earth's history. Its cause is unknown. One possibility is an impact event like that which caused the K-T extinction event (see below). Since an impact event would likely hit an ocean (70 % of the Earth's surface is ocean) , and since the ocean floor replenishes itself approximately every 200 million years due to plate tectonics, it is unlikely that the impact event will still show a crater. The extinction event marks the beginning of the 186 million year-long "Age of Dinosaurs" which is divided, from oldest to most recent, into the Triassic, Jurassic, and Cretaceous periods.

200 million years ago mammals evolved. Mammals have the distinguishing features of mammary glands, sweat glands, and an ability to maintain their internal temperature, i.e. they are warm blooded. This provided a significant evolutionary advantage over cold blooded animals because they could hunt or forage for food during the cold of the night or during the heat of the day. Mammals can lower their temperature by sweating and increase their temperature by shivering. In addition mammals use non-shivering thermogenesis to create heat. The non-shivering thermogenesis is accomplished by accelerating protons across an electrical potential and converting their kinetic energy into heat.

180 million years ago Pangaea began separating. At an average rate of one to two inches per year the continents would eventual end up in their present configuration.

145 million years ago the Cretaceous period began. The name Cretaceous comes from the Latin "creta" meaning chalk. The geologic strata layers of the Cretaceous are characterized by beds of chalk and limestone (calcium carbonate, $CaCO_3$) which were deposited by marine life. The white cliffs of Dover are an example.

80 million years ago the Rockies were formed. As Africa and North America separated following the breakup of Pangaea, and as the North American plate moved northwest, the west coast of North America converged upon and subducted the Farallon plate raising the Rockies (the Laramide orogeny) and depressing central North America. The depressed central region, along with higher sea levels during the late Cretaceous, allowed ocean waters from the Gulf of Mexico and from the Arctic Ocean to flood central North America up to 2500 feet deep (the Western Interior Seaway) which ultimately resulted in limestone deposits hundreds of feet thick. The Western

Interior Seaway, which started in the mid Cretaceous, ended shortly after the end of the Cretaceous period due the continuing uplift of the central regions as the North America plate continued to slide over the Farallon plate. The Farallon plate is named for the Farallon Islands located about 25 miles from San Francisco; farallon means sea cliff in Spanish.

65 million years ago an asteroid killed the dinosaurs. An asteroid approximately six miles in diameter impacted the earth just north of the Yucatan peninsula and formed a crater 100 miles in diameter, the Chicxulub[1] crater. As a result of the impact, particles were ejected into the atmosphere which blocked much of the solar energy for up to 10 years resulting in the extinction of many plants and all large land animals including the dinosaurs. In total 75% of all species became extinct. This was the last of the "Big Five" mass extinctions that occurred since the Cambrian Explosion. Organisms that did not require photosynthesis, such as fungi survived the event. Most fish and amphibians survived. Fortunately, mammals survived leading ultimately to humans and the civilization of mankind around 3300 BC. This mass extinction is often referred to as the K-T extinction event because it occurs at the boundary between the Cretaceous (K) period and the Tertiary (T) period. Note, the name Tertiary dates to 1759 and the International Commission on Stratigraphy has since replaced the Tertiary period by the two periods Paleogene and Neogene. Asteroids have hundreds of times more iridium content than the Earth's crust, and the geologist Walter Alvarez noted that at the K-T boundary

---

[1] The name Chicxulub is the name of a small town on the Yucatan near the impact site (the name means "flea devil" in the local Yucatec Maya language).

the iridium levels were high throughout the world, leading him, in 1980, to the suggestion that an asteroid had impacted the earth causing the mass extinction and the ejected material caused the higher iridium at the K-T boundary throughout the world.

65 million years ago primates evolved. This date varies between 60 MYA and 85 MYA depending on the source. Primates are distinguished from their predecessors by having a larger brain and by an increased reliance on binocular vision. Also, primates have five fingers, five toes, and opposable thumbs. The earliest primates are referred to as prosimians, which means before the simians (monkeys and apes). The prosimians were similar to the modern day Lemur and were about the size of a squirrel. They would be followed, from an evolutionary point of view, by monkeys, gibbons, orangutans, gorillas, chimpanzees, and hominins. Hominins include all relatives of modern humans that are closer than the chimpanzee, for example Homo habilis, Homo erectus, and Neanderthal.

45 million years ago the Himalayas were formed. When the Indian plate separated from Africa it headed north at a speed of about 8 inches per year, the fastest speed known for a tectonic plate, and 45 million years ago it impacted southern Asia pushing up the Himalayas. Today the plate continues to push north at a speed of 2 inches per year, and the Himalayas continue to increase in height by about one inch per year.

37 million years ago monkeys evolved from the prosimians. Over time, some of them made their way, perhaps on floating debris, to Central or South America becoming the New World Monkeys and some remained in Africa and Asia, referred to as the Old World Monkeys. New world monkeys have prehensile tails but Old World monkeys do not. Old world monkeys have trichromatic vision but most New World monkeys do not. Monkeys have larger brains and more forward looking eyes then their prosimian relatives. The monkeys out-competed the prosimians relegating them to nocturnal living in areas away from monkey habitat.

30 million years ago the Alps were formed. The African plate moved northeast at about one inch per year impacting Europe forming the Alps as well as the Carpathians, the Pyrenees, the Apennines, and the Zagros. All of these mountain building events are included in the Alpine orogeny.

18 million years ago apes evolved. Apes evolved from Old World Monkeys. The term "ape" includes gibbons, orangutans, gorillas, chimpanzees, and hominins. Relative to monkeys, apes have a larger brain, have a more upright posture, do not have a tail, and can use simple tools. The figure below shows the skeletal structure of the various apes which are extant today, of which one is the human. It would be hard for anyone to argue that we are not related. The gibbon dates to 18 MYA, the orangutan dates to 14 MYA, the gorilla dates to 7 MYA, and the chimpanzees and hominins date to 5.3 MYA.

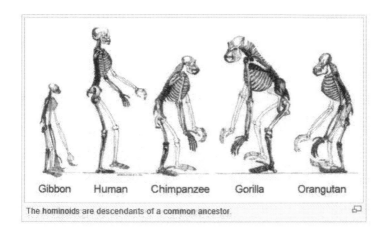

Gibbon   Human   Chimpanzee   Gorilla   Orangutan

The hominoids are descendants of a common ancestor.

5.3 million years ago chimpanzees and hominins speciated from a common ancestor. As mentioned earlier, the designation "hominins" includes modern humans as well as approximately 20 antecedent humanlike species, all extinct, which are closer to humans than the chimpanzee. Note: the word *hominid* used to be defined precisely the same as the word hominin but *hominid* has recently been redefined to include all modern and extinct great apes (i.e.: chimpanzees, orangutans, gorillas and humans). However, at the time of this writing, the scientific community has yet to be on the same page with this change so there is much inconsistency in the literature. Compared to their antecedents, the hominins are characterized as having bipedalism and a larger brain. These features were perhaps encouraged by the spread of grasslands and the evolution of carnivores which occurred at this time. The fossil record of the hominins shows for the first time a pelvis that allows two-legged upright walking; this pelvis design has been passed down, with improvements along the way, to modern humans. In contrast, the chimpanzee pelvis does not allow for significant upright walking, they typically employ knuckle walking.

5.3 million years ago the Mediterranean refilled. After 300,000 years of being dried out after the Straits of Gibraltar closed (due to the gradual northward movement of the African plate). The Mediterranean refilled (referred to as the Zanclean Flood) when the straits reopened due to erosion and/or faulting. Ocean levels dropped 30 ft. as the Mediterranean filled.

2.6 million years ago the Isthmus of Panama was formed. The Caribbean plate collided with the Pacific plate forming the Isthmus of Panama which prevented the circulation of currents between the Atlantic and the Pacific oceans and caused the Gulf Stream to form, changing weather patterns, resulting in the current Ice Age.

2.6 million years ago the current Ice Age began. The term "Ice Age" and "glacial period" are often used synonymously in popular literature. However, an Ice Age typically lasts tens of millions of years and consists of many glacial periods separated by periods of warmer weather. There have been five Ice Ages[1] identified during the last 2.4 billion years. The current Ice Age has consisted of approximately 44 glacial periods; each period lasted

---

[1] There have been five Ice Ages in the geologic history of the Earth: 1) Huronian 2.4 BYA to 2.1 BYA; Cryogenian, aka Snowball Earth, 850 MYA to 630 MYA; Andean-Saharan 460 MYA to 420 MYA; Karoo 360 MYA to 260 MYA; and Current Ice Age 2.6 MYA to today.

initially about 41,000 years and more recently about 100,000 years. The glacial periods are separated by interglacial periods of about 15,000 years. We are currently 11,600 years into the Holocene Interglacial and another glacial period is no doubt in our future however there is no consensus as to precisely when this will begin. The periodicity of the glacial periods is believed to be caused by "Milankovitch cycles" which include variations in the Earths tilt over a 41,000 year cycle (nominally 23.5 degrees but it varies between 22.1 and 24.5 degrees), precession of the Earth's axis of rotation on a 26,000 year cycle, and orbital eccentricity variations on a 100,000 year cycle. The eccentricity varies between 0.000 (a circle) and .06 (currently at .0167). The eccentricity variation is caused by the gravitational effects of Jupiter and Saturn. The average distance from the Earth to the Sun is 93 million miles, but due to the eccentricity the Earth currently is 3 million miles closer in December.

2.6 million years ago Homo habilis evolved. Homo habilis "Handy Man" evolved in Africa and is the first species of the genus Homo, which is characterized by its longer legs, larger brain, and its ability to use stone tools. Also, the evolution of Homo habilis is coincident with the global climate change caused by the formation of the Isthmus of Panama. As part of the climate change the grasslands in Africa expanded, perhaps favoring a species with longer legs and a larger brain that could avoid predators. Homo habilis had a brain that was 30 percent larger than its antecedent but still only 1/3 of current human brain capacity. Homo habilis remained in Africa and for approximately 500,000 years lived alongside the more advanced Homo erectus.

1.9 million years ago Homo erectus evolved. Homo erectus "Upright Man" evolved in Africa with a brain 50% larger than Homo habilis but still only half that of modern humans. Homo erectus could control fire and travelled as far as present day China.

600,000 years ago Neanderthals appeared in Europe. The name "Neanderthal" derives from the Neander Valley in Germany where the first discoveries were made. Neanderthals were adapted for living in cold climates as they existed through five ice ages in Europe. They are genetically very close to modern humans and recent genetic studies have shown that Neanderthal genes were passed to Humans through interbreeding.

200,000 years ago modern humans evolved in Africa. Homo sapiens (modern humans) evolved in Africa at the beginning of the next to the last glacial period. It is speculated that the challenges of survival created by this penultimate ice age led to the collapse of archaic humans and the speciation of modern humans.

70,000 years ago the Toba supervolcano erupted resulting in a human genetic bottleneck. One might think that with the wide range of physical appearances in humans that we are a genetically diverse species, but this is not true, modern humans are much more genetically close than would be expected given our 200,000 year history. One theory as to why we are so genetically close is the Toba Catastrophe Theory. About 70,000 years ago there was a supervolcanic eruption near present day Lake Toba in Indonesia. This eruption caused a global volcanic winter that lasted for up to 10 years. According to the Toba Catastrophe Theory, the aftermath of the eruption reduced the human population to around 15,000 individuals creating a genetic bottleneck. This genetic bottleneck would explain the low genetic diversity observed today in humans.

70,000 years ago modern humans migrated out of Africa. Modern human migrated out of Africa about 70,000 years ago. Generation by generation they gradually filled the habitable world. They reached Southeast Asia 50,000 years ago, Europe 30,000 years ago, and North America 11,600 years ago.

30,000 years ago modern humans arrived in Europe. When modern humans arrived in Europe they encountered the Neanderthals, shortly after which evidence of Neanderthals ceases to exist. Genetic research has concluded that interbreeding between the humans and Neanderthals occurred and modern European and Asian humans have a small percentage of Neanderthal genes.

20,000 years ago the "Last Glacial Maximum" occurred. The *glacial maximum* is when the glaciers were at their maximum extent. In North America ice sheets two miles thick extended south to Long Island New York and to the Ohio River Valley. Long Island was formed by glacial debris being deposited at the glacial terminus where the rate of glacial advance equaled the melt rate; and the Ohio River was shaped by glaciers.

11,000 BC was the start of the Younger Dryas. The Younger Dryas was an abrupt and 1300 year return to glacial temperatures. It was perhaps caused when Lake Agassiz (a glacial lake centered In Canada north of Minnesota and almost twice as large as all the great lakes combined) emptied into the north Atlantic. The influx of the fresh water into the north Atlantic had an immediate effect of prevented the Gulf Stream circulation which in turn caused colder temperatures in Europe, this then resulted in ice sheets advancing again which more permanently (for 1000 years) shut down the Gulf stream until temperatures gradually increased.

9600 BC the last glacial period ended. After the end of the glacial period the ice sheets continued to melt for 4000 years.

9600 BC humans crossed the land bridge from Asia to Alaska. Within 1000 years humans populated North and South America. During this 1000-year period the woolly mammoth and the horse became extinct likely due to hunting by the newly arrived humans. The horse would not set foot on North or South America again until the Spaniards introduced them to Central America after discovery of the "New World" by Christopher Columbus.

9600 BC agriculture began in the "Fertile Crescent". The Fertile Crescent includes Mesopotamia (the land between the Tigris and Euphrates rivers), the Levant (land adjacent to the eastern Mediterranean, and Egypt along the Nile. Crops consisted of wheat and barley. 9600 BC is also the beginning of the Neolithic or new Stone Age, which ends with the beginning of the Bronze Age in 3300 BC, which is also the time for the beginning of civilization.

7000 BC farm animals were domesticated. This included sheep, goats, pigs, and cattle. The horse, a much more difficult animal, was domesticated C4000 BC on the Eurasian steppes, i.e. north of the Caspian Sea.

C5600 BC the Mediterranean overflowed into the Black Sea. Sea levels rose over 400 feet when the ice sheets melted causing the Mediterranean to overflow into the Black Sea (the Black Sea Deluge Theory) which was several hundred feet lower. The volume was 200 times that of the Niagara Falls, permanently flooding thousands of square miles of land with a resulting migration of many people. This is considered by some to be the source for the great flood described in the Bible, the Quran, and the Epic of Gilgamesh.

5300 BC the first settlements appeared in Mesopotamia and Egypt. The word *settlement* means a gathering together of people into a single geographic area. As an example, a settlement may contain 1000 people spread over 10 acres residing in 200 small houses (5 persons per house). A settlement would include family housing and common buildings. Surrounding the living areas would be farm land and cattle grazing land as well as a nearby water source such as a river or lake.

C3500 BC the wheel was invented. This early wheel did not have spokes. Spoked wheels and the chariot were invented c2000 BC and used effectively by the Hittites when they settled in the Anatolian peninsula.

3300 BC civilization began in Mesopotamia and in Egypt. The date for the start of civilization varies widely depending on numerous definitions and assumptions. The date of 3300 BC was chosen by the author since by that point in time developments included a central government, religion, written language in the form of cuneiform in Babylonia and hieroglyphs in Egypt, and the basics of measurement science. It is also considered as the beginning of the Bronze Age. Not addressed in this book, but also about this time, civilization was beginning in India, China, and the Americas.

The Fertile Crescent

Geography

The Fertile Crescent is where early civilization started in the Middle East. This includes Egypt, the Levant, and Mesopotamia as shown here.

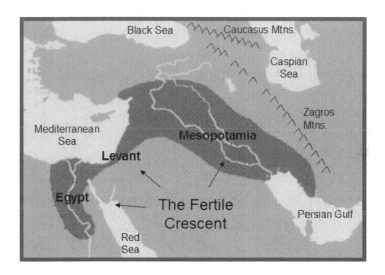

Humans have lived in the Fertile Crescent as hunter-gatherers for tens of thousands of years, however, it is only since the end of the last ice age (9600 BC) that animals and wheat were domesticated and humans started down a path to civilization. The impetus toward civilization was provided by two migration events. These migrations resulted in more people in the Fertile Crescent and c5000 BC there were numerous farming communities using Stone Age tools growing wheat and raising domesticated sheep and goats. Around 3300 BC "civilization[1]" began which included the invention of writing followed by the development of number systems and standards of measurement.

Migrations

Migrations from the Sahara
During the last ice age, a period of 100,000 years, the Sahara desert was unsuitable for human habitation; it was desert like it is today. However, c9600 BC the two mile thick ice sheets over northern Europe started to melt, marking the end of the ice age. The melting ice sheets, which took 4000 years for the melt, caused climatic changes which had far reaching effects including bringing rain to the Sahara allowing human habitation, at the end of the melt c5600 BC the Sahara returned to desert causing human migration to the Fertile Crescent.

Migrations from the Black Sea Area
As mentioned earlier, the melting ice sheets caused an over 400 feet rise in ocean levels with the result that, c5600 BC, the Mediterranean overflowed into the Black Sea permanently flooded thousands of square miles of land with a resulting migration of many peoples. It has been speculated that this is the source for the Biblical Great Flood story.

---

[1] There is not an agreed upon definition for the word civilization but it would be generally agreed that a population that exhibited the following traits would be considered civilized: a central government, an established religion, means for writing and for recording information, standards of measurement, rules for mathematical calculations, division of labor, trade with neighboring regions, and laws.

Migrations from the Pontic Steppe

Indo-European speakers were located in the area of the Pontic steppe which is a grassland plane as large as Texas and New Mexico combined; it is located north of the Black Sea and Caspian Sea as shown here. The name Pontic derives from the area on the south coast of the Black Sea which was known by the ancient Greeks as the *Pontos Euxeinos* or Hospitality Sea.

Pontic Steppe

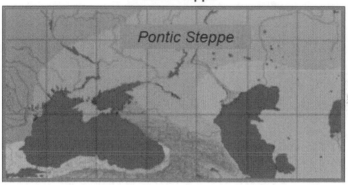

Horses were first domesticated in this area c4000 BC. Migrations southward into our region of interest occurred well after the beginning of civilization. These migrations occurred in several stages. C2000 BC the Hittites migrated into Anatolia where they developed iron smelting but kept the process a secret. After the Thera eruption of 1628 BC the Dorians and Ionians migrated into what is now Greece. C1200 BC was the "Bronze Age Collapse" and the invasion of the Sea People causing turmoil in the eastern Mediterranean and much movement of peoples. This was the beginning of the Greek Dark Ages and the Babylonian Dark Ages, and very close in time, considering the uncertainties in the ancient timeline, to when the Jews left Egypt. The origin of the Sea People remains a mystery to this day. As part of the Bronze Age collapse the Hittite Empire ceased to exist with the result that their iron processing technology spread throughout the Fertile Crescent marking the beginning of the Iron Age.

Languages

Ancient Languages or language subfamilies within the Fertile Crescent were Sumerian, Elamite, Egyptian, Semitic, and eventually Indo-European. These languages are discussed here.

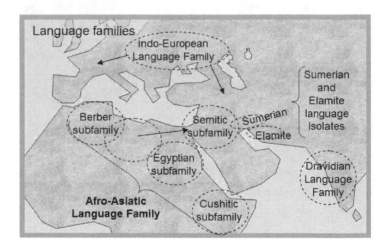

Egyptian Subfamily

The Ancient Egyptian language is part of the Egyptian subfamily, one of four subfamilies of the Afro-Asiatic Family of languages. It was recorded c3400 BC making it one of the oldest languages. It was spoken for 5000 years and was gradually replaced by Egyptian Arabic after the Muslin conquests c700 AD.

Sumerian and Elamite Languages
In the Tigris-Euphrates River delta region lived the Sumerians and nearby along the north side of the Persian Gulf were the Elamites. The Sumerian and Elamite languages are considered language isolates since they have not been strongly tied to any language family or to each other. It has been proposed by linguist David McAlpin that Elamite and the Dravidian languages (Dravidian is spoken in southern India) are part of an Elamo-Dravidian language family, and that Harappan (spoken in the Indus Valley) may share this background. The Sumerians invented agriculture, irrigation, the potter's wheel, and cuneiform writing on clay tablets of which hundreds of thousands exist. On their farms they harvested wheat and barley and raised sheep, goats and pigs, all of which had been domesticated by 7000 BC. Sumerian has only been deciphered in the last hundred years.

Semitic languages
Semitic people from northern Africa settled in the lands adjacent to the eastern Mediterranean known as the Levant (from French for the Orient) and in the Tigris-Euphrates regions north of Sumeria. Over the millennia the Semitic language diverged as shown in the language family tree. The Phoenicians on the Mediterranean coast invented c1100 BC an alphabet based on consonants, no vowels. This alphabet was used with some variation for writing in Aramaic and Hebrew. Aramaic (spoken by Jesus) was originally spoken in the region near modern Syria.

Akkadian was spoken in central Mesopotamia. Akkadian spawned Assyrian and Babylonian.

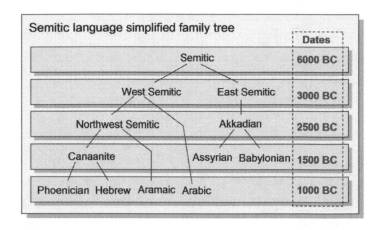

Indo-European languages
People speaking Indo-European languages migrated down from the north into the bordering regions of Mesopotamia starting c2000 BC. Indo-European is the root language of most languages in Europe with a few exceptions. Basque, located in the Pyrenees, is a language isolate. Finnish, Hungarian and Estonian are from the Uralic language family (speculated to have originated near the Ural Mountains). Turkish is from the Turkic language family. Lastly, Maltese, spoken on Malta, is the only Semitic language of national status in Europe.

Egypt

Geography

Egypt is located in the northeast corner of Africa and is about as large as France and Germany combined. It is mostly hot arid desert except for areas along the Nile.

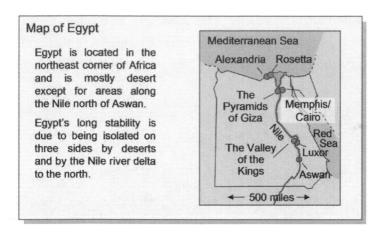

Map of Egypt

Egypt is located in the northeast corner of Africa and is mostly desert except for areas along the Nile north of Aswan.

Egypt's long stability is due to being isolated on three sides by deserts and by the Nile river delta to the north.

Mediterranean Sea

Alexandria Rosetta

The Pyramids of Giza

Memphis/ Cairo

The Valley of the Kings

Red Sea

Luxor

Aswan

◄— 500 miles —►

The Nile delta is about the size of Massachusetts. Every year, thousands of miles to the south, the spring rains in Ethiopia caused the Nile in Egypt to overflow its banks and flood the adjacent lands, referred to as the inundation, bringing nutrients to the soil which are needed for a reliable agriculture; this was the situation in Ancient Egypt, since then, the Aswan High Dam, completed in 1976, controls the river flow and flooding. Because of the deserts located to the east, west, and south, and because of the marshy Nile delta to the north, Egypt was effectively isolated from other civilizations which contributed, along with its reliable agriculture, and along with its religion (to be discussed later) to a very stable society. This is in direct contrast to Babylon which was at the crossroads of world trade.

Ancient Egyptian Timeline
The Ancient Egyptian timeline has been divided by historians into nine periods, chief among them being the Old Kingdom, the Middle Kingdom, and the New Kingdom, which begin approximately 2700 BC, 2100 BC, and 1600 BC. These and some of the other periods are discussed below.

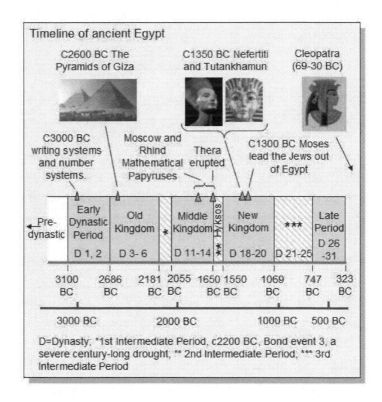

Early Dynastic Period (3100-2686 BC)

Upper Egypt and Lower Egypt were united by King Menes c3100 BC. He and his descendants formed the First Dynasty of Egypt. The subsequent 25 dynasties ruled Egypt until the Persian conquest in 525 BC.

Old Kingdom (2686-2181 BC)

The Old Kingdom of Egypt is also known as the Age of the Pyramids which started with the step pyramid at Saqqara (about 20 miles south of modern day Cairo), approximately 200 feet tall and originally clad in polished white limestone; it was built under the guidance of the famous vizier Imhotep. Shortly after the Saqqara pyramid, the three great pyramids of Giza, shown here, were constructed for Pharaoh Khufu, his son, and grandson. As a point of interest, also about this time, Stonehenge was built in England.

The Pyramids of Giza, built c2500 BC

First Intermediate Period (2181-2055 BC)

The Old Kingdom ended when a century long drought stopped the annual flooding of the Nile bringing economic ruin with the result that governmental control was transferred to the regional governors. This drought is referred to as Bond event 3. Gerald C. Bond (1940-2005 AD), American geologist, speculated that there were periodic climatic events approximately every 1500 years, of which this is one. There are a total of nine Bond events identified dating back to 9100 BC.

Middle Kingdom (2055-1650 BC)

When the century long drought which caused the end of the Old Kingdom ended, Egypt was again unified under one government marking the beginning of the Middle Kingdom. Trade, arts and literature flourished. The two most famous mathematical papyruses, the Moscow Papyrus and the Rhind Papyrus, which contain procedures for solving numerous problems, date to the Middle Kingdom. The Middle Kingdom came to an end with the eruption of Thera in 1628 BC. Thera, a volcano located near modern day Santorini in the Aegean, erupted with such force that it destroyed the Minoan civilization on Crete and devastated the regions around the Aegean and the eastern Mediterranean. One of the regions seriously impacted was the Nile delta which was likely hit with a Tsunami about an hour after the eruption.

Second Intermediate Period – Hyksos Rule (1650-1550 BC)

In the aftermath of the Thera eruption the Hyksos took over control of the Egyptian Nile delta (Lower Egypt). The Hyksos are believed to be a Semitic people from the Levant and perhaps further north; they introduced the horse and chariot to Egypt. After 100 years of rule the Hyksos were conquered by pharaoh Ahmose I of Upper Egypt.

New Kingdom (1550-1059 BC)

Reunification (1550 BC)
The New Kingdom began with the reunification of Upper and Lower Egypt after the Hyksos were expelled by Ahmose I, founder of the 18th dynasty. Famous 18th dynasty leaders include Hatshepsut, Akhenaten, Nefertiti, and Tutankhamun.

Hatshepsut (c1450 BC)
Pharaoh Hatshepsut, c1450 BC, the fifth ruler of the 18th dynasty, is regarded as "the first great woman in history of whom we are informed" and is considered to be one of the most successful pharaohs. Her mortuary temple, shown below, located near the Valley of the Kings, is considered one of the "incomparable monuments of ancient Egypt."

Mortuary Temple of Hatshepsut c1450 BC

Akhenaten (c1350 BC)

Akhenaten is famous for his attempt to convert Egypt from polytheism to monotheism. His queen was Nefertiti and his son (not by Nefertiti) was Tutankhamun. Some have speculated that Akhenaten is the Biblical Moses.

Akhenaten

Nefertiti (c1350 BC)

Nefertiti, queen to Akhenaten, is considered to be one of the most beautiful women in history, to which this bust, by sculptor Thutmose, the official sculptor of Pharaoh Akhenaten, will attest; this bust is located at Berlin's Neues Museum.

Nefertiti Bust c1350 BC

Tutankhamun (c1330 BC)

Tutankhamun is the famous pharaoh son of Akhenaten (not by Nefertiti). His undisturbed tomb was discovered in 1922 by Howard Carter (1874-1939 AD), English archeologist.

Burial Mask of Tutankhamen c1330 BC

Moses (c1300 BC)

In the middle of the New Kingdom, c1300 BC, after 200 years in Egypt and after ten plagues, the last of which managed to "passover" the Jews, Moses lead 600,000 Israelites in their great exodus out of Egyptian slavery into the land of Canaan. Canaan consists of the land between the Mediterranean Sea and the Jordan River as well as equal amounts of land north, south, and east of this area. After the exodus, the Ten Commandments were given to Moses on Mount Sinai and according to tradition, Moses wrote the first five books (The Torah) of the Bible (Genesis, Exodus, Leviticus, Numbers, and Deuteronomy). What has been written above in this paragraph is consistent with the Bible and with rabbinic tradition. However, the existence of these events and the associated timeline are the subject of much scholarly study and dispute. For example relative to the Torah, the consensus is that the Torah was taken from four different written sources, dating from 950 BC to 500 BC, and that they were brought together c400 BC, long after the time of Moses.

Ramesses the Great (c1250 BC)

Not too long after the time of Moses was the reign of Pharaoh Ramesses the Great, the third pharaoh of the 19th dynasty, who reigned for 66 years. The temples at Abu Simbel were constructed during his reign. When the Aswan dam was built in the 1960s, the Abu Simbel temple was relocated to higher ground so that it would not be submerged by Lake Nasser.

Temple at Abu Simbel c1250 BC

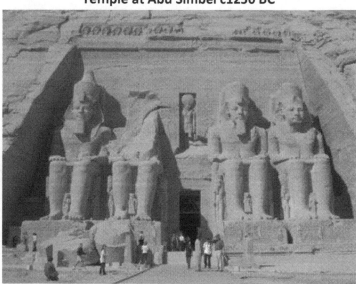

Karnak (c1250 BC)

The Temple of Karnak was built in this period, and the Valley of the Kings, across the Nile from Karnak, was the location of the burial tombs for the pharaohs and nobles.

Temple at Karnak c1250 BC

Third Intermediate Period (1069-747 BC)

The Third Intermediate Period, marked by a dissipation of central power, resulted from several events: uncertainty in the transition of power from Ramesses III (the last and great Pharaoh of the 20th Dynasty); from famines caused by the eruption of the Hekla volcano in Iceland; and from wars with invaders from Libya and the "Sea People". Also, the start of the Third Intermediate Period is close to an event known as the "Bronze Age Collapse" (c1200 BC) which marked the beginning of the Greek Dark Ages, the collapse of the Hittite Empire in Anatolia, the beginning of the Babylonian Dark Ages, and the destruction of major cities in the eastern Mediterranean, including Troy. C1200 BC was not a good period of time in the eastern Mediterranean.

Late Period (747-323 BC)

The last pharaoh of Egypt was Cleopatra, a Greek queen. She took her life in 30 BC as did Mark Anthony when they were defeated by the forces of Caesar Augustus (nephew of Julius Caesar); Augustus went on to become the first Emperor of Rome.

Agriculture

The reliability of Egypt's agriculture was due to fertile land along the Nile which was annually replenished with nutrients when the Nile flooded. After the annual inundation, farms would be resurveyed as necessary and planted with corn, barley, and wheat which would be harvested prior to the next inundation. When Sirius, the Dog Star, was visible in the early morning eastern sky, just prior to the rising of the sun, the Egyptians knew the inundation would soon arrive.

Government

The ruler of Egypt was the pharaoh who was considered to be the son of the sun God Ra. Reporting to the pharaoh was the vizier, who was responsible for running the country, the treasurer, the tax collector, the minister of public works, and the commander of the army. The ministers and advisors to the pharaoh were typically priests who formed a class of nobility. The peasants, who formed the majority of the population, did not own anything and had no say in the government; they paid taxes in goods or labor. The country was divided into 42 nomes which were administered my governors (nomarchs) who reported to the vizier. On average, each nome was about 150 square miles and populated by 20,000 people.

Technology

Papyrus
As early as 3000 BC the Egyptians invented papyrus scrolls made from the papyrus plant, common in the Nile delta; the word paper comes from papyrus. From Wikipedia: *"Papyrus is made from the stem of the papyrus plant. The outer rind is removed and the sticky fibrous inner pith is cut lengthwise into thin strips of about sixteen inches long. The strips are placed on a hard surface with their edges slightly overlapping, and then another layer of strips is laid on top at a right angle. While still moist, the two layers are hammered together, mashing the layers into a single sheet. The sheet is then dried under pressure. After drying, the sheet is polished with a rounded object."* Papyrus sheets were glued together to form scrolls. Around 300 BC the Library of Alexandria was founded in memory of Alexander the Great who died in 323 BC. The library started buying up the production of papyrus to make copies of all known literature resulting in a papyrus shortage; as a result of the shortage, parchment, made from calf, sheep, or goat skin, became a popular substitute.

Papyrus Plant

Hieroglyphs

C3000 BC the Egyptians invented hieroglyphs for writing which used over two thousand different pictograms. It would be another two thousand years (c1050 BC) before the Phoenicians, who flourished along the coast of the eastern Mediterranean near modern day Lebanon, developed a twenty two letter alphabet consisting of consonants only. The Greeks added vowels c850 BC allowing Homer c800 BC to better express himself in the Iliad and the Odyssey.

Hieroglyphs Written on Papyrus

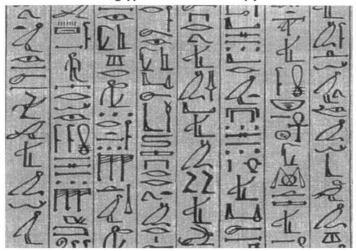

Ship Building

C2400 BC the Egyptians made ships that could sail into the wind by tacking. Note the rope truss down the middle of the ship to give the hull rigidity. This was 800 years before the Phoenicians started sailing the Mediterranean.

Ancient Egyptian Ship

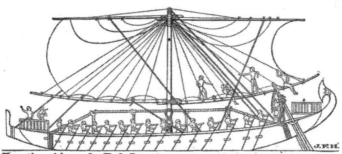

Egyptian ship on the Red Sea, about 1250 B.C. [From Torr's "Ancient Ships.")
Mr. Langton Cole calls attention to the rope truss in this illustration, stiffening the beam

Mathematics

C3000 BC the Egyptians developed cookbook procedures for measurement, geometry, and mathematics. They discovered that a 3-4-5- triangle formed a right triangle but there is no indication that they understood the Pythagorean Theorem. The procedures provided answers of practical utility but they never extended beyond their immediate need and they were not developed from fundamental concepts; it would be another two thousand years before the enlightened Greeks took that step. One of the Egyptian's famous formulas showed how to calculate volume of a truncated pyramid; how this might have been accomplished is addressed later in this book.

Religion

Ancient Egyptians believed that in the beginning there was chaos from which the sun god Ra created all life and from which his daughter, the goddess Maat, created order, justice, and harmony. Pharaohs are frequently shown making an offering to Maat, showing that they are preserving the principles of Maat on earth. The Pharaoh was believed to be descended from the gods. He acted as the intermediary between his people and the gods, and was obligated to sustain the gods through rituals and offerings so that they could maintain order in the universe. The state dedicated enormous resources to rituals and to the construction of temples.

The concept of Maat (order, justice and harmony) pervaded Egyptian life. Maat bound the universe, the natural world, the state, and the individual in an indestructible unity. Cosmic harmony was achieved by correct public and ritual life. Maat emphasized adherence to tradition as opposed to change. Any disturbance in cosmic harmony could have consequences for the individual as well as the state. An impious King could bring about famine; or individual blasphemy could cause blindness.

In opposition to Maat was a concept known as Isfet which was disorder, injustice, and imbalance. Life was a struggle between Maat and Isfet.

Ancient Egyptians believed in a soul and an afterlife; they made great efforts to ensure the survival of their souls after death, providing tombs, grave goods, and offerings to preserve the bodies and spirits of the deceased. The soul was composed of five parts: heart, shadow, name, personality, and vital essence. The heart was the most important part of the soul. Good deeds made your heart lighter and bad deeds made your heart heavier. At death your soul would enter the underworld, a vast region known as Duat, where your heart would be weighed by Anubis, god of the afterlife, and if your heart is light enough you would go to heaven, if your heart is too heavy, from bad deeds, your heart would be devoured by a demon and you would forevermore cease to exist.

As can be surmised from the above discussion ancient Egypt had a polytheistic religion. Akhenaten, a pharaoh of the 18th dynasty, his queen was Nefertiti, tried to discard the traditional polytheism and change to a theology based on a single God. As mentioned earlier, since his rule was coincident with the time of Moses; it has been speculated that he obtained this views from the Jews, or, some also speculate that Moses and Akhenaten were the same person.

Later Egyptian Timeline (Foreign Rule)

In 525 BC the Persians took control of Egypt, they would be replaced by the Greeks in 323 BC and by the Romans in 146 AD. When the Western Roman Empire fell to the Germanic tribes in 476 AD control of Egypt passed to the Eastern Roman Empire, i.e., Byzantium.

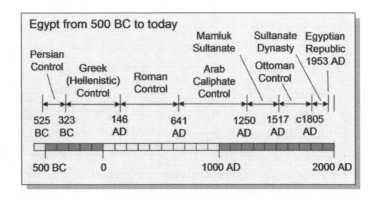

Mohammed founded Islam c620 AD and in 641 AD Egypt came under Arab Caliphate control which lasted for approximately 600 years. In 1250 AD control passed to the Mamluks. The Mamluks were a slave/warrior caste from regions north of the Black Sea whose services were purchased by the Arab ruling class to defend their caliphates. Under Mamluk leadership, and encouraged by the fall of Baghdad to the grandson of Genghis Khan in 1258 AD[1], Cairo became the center of Islamic learning.

---

[1] Circa 1220 AD Genghis Kahn was insulted by the Shah of Iran, provoking him to travel 2000 miles and attack Persia which resulted in the sacking of the capital Samarkand and the defeat of the Shah. The Kahn's grandson expanded the attack when he conquered Mesopotamia and sacked Baghdad in 1258 AD.

Mamluk control lasted until the Ottoman Empire conquered Egypt in 1517 AD. Napoleon Bonaparte invaded Egypt in 1798 but was repulsed by the combined Ottoman, Mamluk, and British forces in 1801. One of Napoleon's soldiers discovered the Rosetta stone which provided the clues necessary to allow scholars to translate ancient Egyptian hieroglyphs. Occupation by Napoleon was followed by four years of struggle for Egyptian control between Albania, the Mamluks, and the Ottomans. In 1805 the Albanian commander Mohammad Ali became Sultan of Egypt establishing a dynasty that would last until 1953 AD. In 1953 AD, at long last, Egypt returned to Egyptian, albeit authoritarian, control.

Egyptian Mathematics

The Rhind and Moscow Mathematical Papyruses

Egyptian mathematics was based on cook book procedures that met the needs for record keeping, surveying, construction, and tax calculation. How the procedures were developed is not known but they were recorded in the Rhind and Moscow Mathematical Papyruses. The Rhind Mathematical Papyrus is an Egyptian scroll dated to 1650 BC; when unrolled it is 13 inches high by 16 feet long. It contains a list of 84 mathematical problems related to basic arithmetic, the calculation of areas and volume, the calculation of the slope of a pyramid, and the calculation of recipe proportions. The papyrus was purchased in 1858 in Luxor Egypt by Alexander Henry Rhind, a Scottish Egyptologist, hence the name. The Moscow Mathematical Papyrus is 200 years older than the Rhind papyrus and resides in the Pushkin State Museum of Fine Arts in Moscow. It contains solutions to 25 problems including a formula to calculate the volume if a truncated pyramid.

The Cubit

Egyptian standards of measurement date to 3000 BC and the unit of length was the cubit. There were two types of cubits, the Standard Cubit and the Royal Cubit.

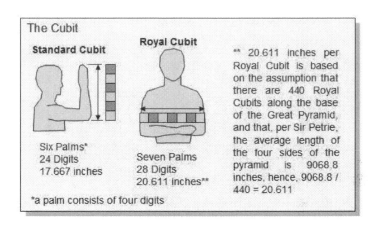

The Cubit

Standard Cubit

Royal Cubit

Six Palms*
24 Digits
17.667 inches

Seven Palms
28 Digits
20.611 inches**

** 20.611 inches per Royal Cubit is based on the assumption that there are 440 Royal Cubits along the base of the Great Pyramid, and that, per Sir Petrie, the average length of the four sides of the pyramid is 9068.8 inches, hence, 9068.8 / 440 = 20.611

*a palm consists of four digits

A Standard Cubit was the distance from the elbow to the finger tips; it was divided into six palms with four fingers each. Note that the "palm" does not include the thumb. The Royal Cubit (20.611 inches) was seven palms of four fingers each and was measured based on the elbow to elbow distance when the pharaoh was in his royal pose. The Cubit was not standardized like measurements today; it would vary with location and time. At the time the Great Pyramid was built the Royal cubit is estimated to be 20.611 inches. As noted by Sir Petrie, there is evidence that the Royal Cubit was sometimes divided into ten parts.

The Seked

The Seked is an Egyptian measurement of the slope of a surface much like the pitch of a roof is measured today. Roof pitch is the amount of rise (vertical distance) given a one foot run (horizontal distance). For example, a seven pitch is seven inches of rise for one foot of run.

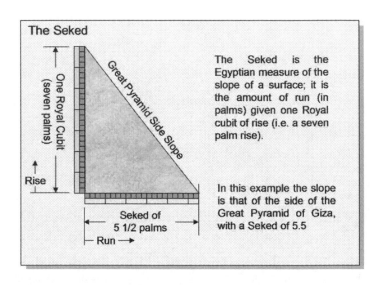

The Seked is the Egyptian measure of the slope of a surface; it is the amount of run (in palms) given one Royal cubit of rise (i.e. a seven palm rise).

In this example the slope is that of the side of the Great Pyramid of Giza, with a Seked of 5.5

The Egyptians used a similar but reverse ratio, they measured slope as the amount of run in "palms" for a one cubit (i.e. seven palms) rise. The Great Pyramid has a height of 280 cubits and distance of 220 cubits from the center of a side of the square base to a point under the apex of the pyramid. This gives side slope of 220 run to 280 rise; or, dividing by 40 gives 5.5 run to 7 rise; since a cubit has seven palms, the Seked of the Great Pyramid is then 5.5.

Egyptian Fractions

Egyptian fractions were developed during the Middle Kingdom. The Egyptians used unit fractions i.e., fractions with a one in the numerator; as an example, the number ¾ would be represented as ½ + ¼. An exception was the fraction 2/3 for which a unique symbol was used.

To Multiply by Adding

The Egyptians multiplied by adding. When two numbers are multiplied, the first number is the *multiplier* and the second is the *multiplicand*, we will use these names below. A *series* is a sequence of numbers, we will need the following series for this discussion: 1, 2, 4, 8, 16, 32, 64, etc. note that each number is twice its preceding number; any number can be represented as the sum of numbers in this series, for example 13 equals 1 + 4 + 8.

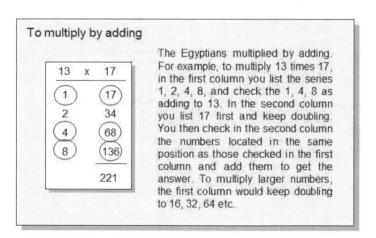

To multiply by adding

The Egyptians multiplied by adding. For example, to multiply 13 times 17, in the first column you list the series 1, 2, 4, 8, and check the 1, 4, 8 as adding to 13. In the second column you list 17 first and keep doubling. You then check in the second column the numbers located in the same position as those checked in the first column and add them to get the answer. To multiply larger numbers, the first column would keep doubling to 16, 32, 64 etc.

The Egyptians multiplied using a doubling table. In the first column they would list the series 1, 2, 4, 8, etc. as necessary such that the multiplier can be represented by the sum of selected numbers in the series. In the second column they would write the multiplicand alongside the 1 in the first column and then keep doubling it. They would then mark the numbers in the first column that add to the multiplier and mark the corresponding numbers in the multiplicand column. If you add the marked numbers in the multiplicand column the result is the desires multiplication.

The Area of a Parcel of Land

The Egyptians calculated land area to levy taxes. They knew that the area of a rectangle or parallelogram is equal to the length of its base times the length its height; and that the area of a triangle was equal to ½ the base times the height. They would divide a complicated area into simpler areas.

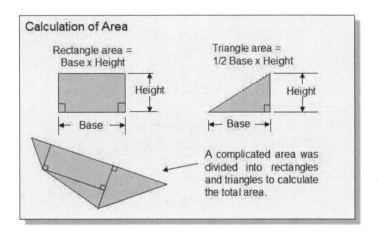

Determining a Right Angle

Egyptian surveyors were referred to as rope stretchers because they used long ropes marked off in cubits for measuring land. A right angle was established by laying out sections of rope in a ratio of 3-4-5 as shown here.

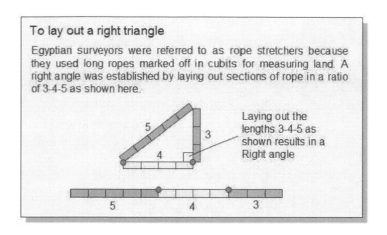
The Area of a Circle

The Egyptians estimated the area of a circle as being equal to the area of a square with sides 8/9 the diameter of the circle. For example, if the circle has a diameter of 9, then the related square has a side of 8 (i.e., 8/9 x 9 = 8), and therefore the area of the square is 64, which is also the Egyptian estimate of the area of the circle.

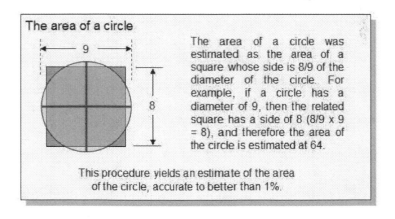

This procedure yields an excellent estimate, accurate to better than 1%. The actual area is Pi x r^2 (we will derive this equation when we study Archimedes) where r is the radius of the circle and Pi is the ratio of the circumference of a circle to its diameter, a value approximately equal to 3.1415926. So for this example the actual area is 3.1415926 x 4.5 x 4.5 = 63.617 versus the Egyptian estimate of 64.

The Volume of a Square Tower or Cylindrical Granary

The Egyptians calculated the volume of a column or a cylindrical granary (where grain is stored) exactly as we do today by calculating the area of the base and multiplying by the height.

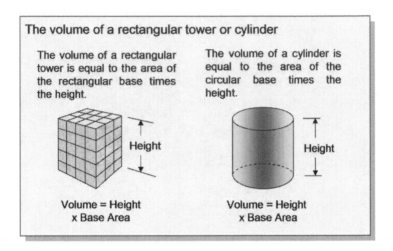

The volume of a rectangular tower or cylinder

The volume of a rectangular tower is equal to the area of the rectangular base times the height.

The volume of a cylinder is equal to the area of the circular base times the height.

Height

Height

Volume = Height x Base Area

Volume = Height x Base Area

The Volume of a Pyramid -- Equation

The Moscow Mathematical Papyrus provides the Egyptian procedure for calculating the volume of a truncated pyramid.

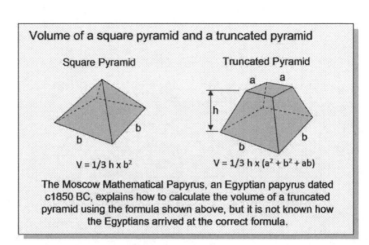

Volume of a square pyramid and a truncated pyramid

Square Pyramid

Truncated Pyramid

$V = 1/3\ h \times b^2$

$V = 1/3\ h \times (a^2 + b^2 + ab)$

The Moscow Mathematical Papyrus, an Egyptian papyrus dated c1850 BC, explains how to calculate the volume of a truncated pyramid using the formula shown above, but it is not known how the Egyptians arrived at the correct formula.

Let "a" be the length of a side of the square truncated top, and let "b" be the length of a side of the square bottom, and let "h" be the height, then the volume (V) of the truncated pyramid is: $V = 1/3\ h \times (a^2 + b^2 + ab)$.

If the truncated piece is zero, i.e. "a" equals zero, then we have the formula for the volume of a regular pyramid.

$V = 1/3\ h \times b^2$

It is not known how the Egyptians came up with the formula for a truncated pyramid since they did not have the benefit of algebra or sophisticated geometry techniques. However, a possible approach is discussed in the following paragraphs.

Pyramid Defined

A pyramid is often thought of as a solid with a square base, four sides, and a centered pointed top. The definition here is more general in that it allows a flat base of any shape and the apex need not be centered above the base. The height of the pyramid is the perpendicular distance between the apex and the plane containing the base.

Definition of a pyramid

A pyramid has a flat base of any shape with straight lines from the perimeter of the base to the apex. The height of a pyramid is the perpendicular distance between the apex and the plane containing the base.

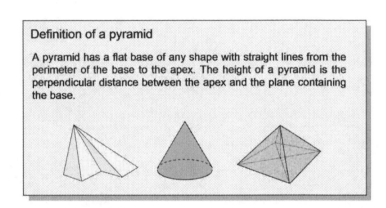

Dividing a Cube into Pyramids

Following the above definition, a cube can be divided into three identical pyramids. Since the volume of a cube is equal to its base area times its height, and since we have divided the cube into three equal pyramids, then the volume of each pyramid is equal to 1/3 the base area times its height.

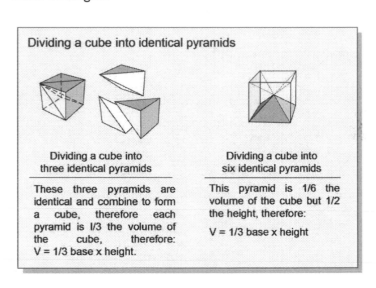

Dividing a cube into identical pyramids

Dividing a cube into three identical pyramids	Dividing a cube into six identical pyramids
These three pyramids are identical and combine to form a cube, therefore each pyramid is 1/3 the volume of the cube, therefore: V = 1/3 base x height.	This pyramid is 1/6 the volume of the cube but 1/2 the height, therefore: V = 1/3 base x height

A cube can also be divided into six equal pyramids (each face of the cube being the base of a pyramid whose apex is at the center of the cube). Since we have twice as many pyramids but their height is half the height of the cube, then the equation V = 1/3 base area x height remains valid. This equation for the volume of a pyramid can be shown to be true in general regardless of: the position of the apex, the height of the pyramid, or the shape of the base.

Properties of a Pyramid

If you consider a step pyramid made of blocks, you can easily see that the volume of the pyramid is independent of skew. I.e., the number of blocks and therefore the volume is constant as you move the apex. If we make the blocks smaller and smaller, we can approach as close as we please a pyramid with smooth sides, therefore, any conclusions we draw based on the step pyramid are also true for a smooth-sided pyramid. Therefore, pyramid volume is independent of skew.

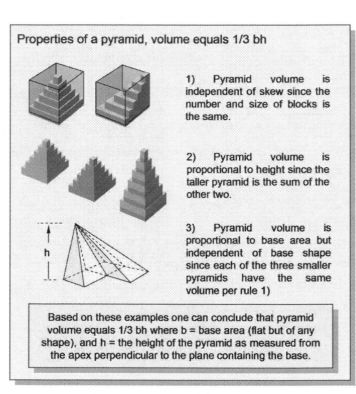

Properties of a pyramid, volume equals 1/3 bh

1) Pyramid volume is independent of skew since the number and size of blocks is the same.

2) Pyramid volume is proportional to height since the taller pyramid is the sum of the other two.

3) Pyramid volume is proportional to base area but independent of base shape since each of the three smaller pyramids have the same volume per rule 1)

Based on these examples one can conclude that pyramid volume equals 1/3 bh where b = base area (flat but of any shape), and h = the height of the pyramid as measured from the apex perpendicular to the plane containing the base.

Similarly, if we add two identical pyramids together to form a taller pyramid, the taller pyramid has the same base area but is twice as high as the two smaller pyramids and has twice the volume. You can generalize this observation to conclude that the pyramid volume is proportional to height. Lastly, if you have a pyramid that has an irregularly shaped base, you can always divide the base into small equal squares forming many smaller pyramids. The volume of a smaller pyramid is equal to the area of the smaller base times the height times 1/3, which is the same for all the smaller pyramids. Therefore, the total volume is equal to the height times the sum of all the smaller square areas, which is the area of the irregular base, times 1/3. Therefore, we conclude that pyramid volume is independent of base shape, and is only dependent on the total base area. To summarize: V = 1/3 base area x height, regardless of the position of the apex, or the shape of the base.

The Volume of a Truncated Pyramid

A truncated pyramid is a regular pyramid but with the top removed along a plane parallel to the base.

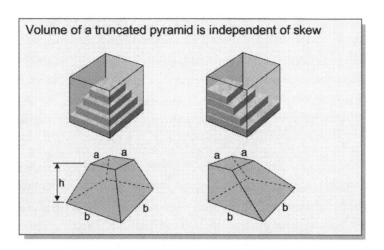

Volume of a truncated pyramid is independent of skew

If the truncated pyramid is skewed as shown it can then conveniently be divided into four smaller pyramids from which the equation for a truncated pyramid can readily be established. Referring to the illustration below, if "a" is the length of a side of the top and "b" is the length of a side of the base, you can see that the truncated pyramid can be divided into four volumes: $V_1 = 1/3$ h x a^2; $V_2 = 1/3$ h x b^2; $V_3 = 1/3$ a x ½ bh; $V_4 = 1/3$ a x ½ bh, and, $V_3 + V_4 = 1/3$ h x ab; therefore $V = 1/3$ h $(a^2 + b^2 + ab)$

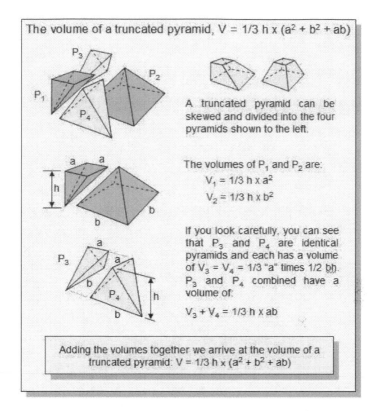

The volume of a truncated pyramid, $V = 1/3\ h \times (a^2 + b^2 + ab)$

A truncated pyramid can be skewed and divided into the four pyramids shown to the left.

The volumes of P_1 and P_2 are:

$$V_1 = 1/3\ h \times a^2$$
$$V_2 = 1/3\ h \times b^2$$

If you look carefully, you can see that P_3 and P_4 are identical pyramids and each has a volume of $V_3 = V_4 = 1/3$ "a" times $1/2\ bh$. P_3 and P_4 combined have a volume of:

$$V_3 + V_4 = 1/3\ h \times ab$$

Adding the volumes together we arrive at the volume of a truncated pyramid: $V = 1/3\ h \times (a^2 + b^2 + ab)$

The ancient Egyptians were every bit as intelligent as people today and there is no reason to believe that they couldn't have gone through this same logic to come up with this formula.

The Great Pyramid of Giza

Introduction

Eight miles southwest of Cairo, on the west bank of the river Nile, is the Giza Plateau upon which sit three large similarly shaped pyramids. The largest is referred to as the Great Pyramid of Giza, the only remaining wonder from the Seven Wonders of the Ancient World[1]. At 481 feet tall it was the tallest structure in the world for 4000 years, finally being exceeded in height by the great cathedrals of Europe built c1400 AD.

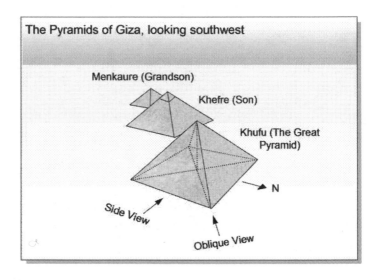

---

[1] The Seven Wonders of the Ancient World was a list of locations for tourists to visit c100 BC. The wonders were: The Great Pyramid of Giza; the Hanging Gardens of Babylon; the Temple of Artemis at Ephesus; the Statue of Zeus at Olympia; the Mausoleum of Halicarnassus; the Colossus of Rhodes; and, the Lighthouse of Alexandria.

The three Pyramids of Giza were constructed c2600 BC. In this timeframe the Egyptians were quite accomplished at design and construction. Using copper saw blades studded with hard stone they could quickly saw through the softer limestone used to make the Great Pyramid. Of the three large pyramids on the Giza Plateau, the first to be built was the Great Pyramid which was finished in 2580 BC after 20 years of construction. It was intended as the final resting place for the Pharaoh Khufu (also referred to as Cheops by Herodotus). Khufu was the 17th Pharaoh to rule Egypt since the first Pharaoh Menes united North and South Egypt in 3100 BC. The other two pyramids on the Giza Plateau were for Khufu's son and grandson.

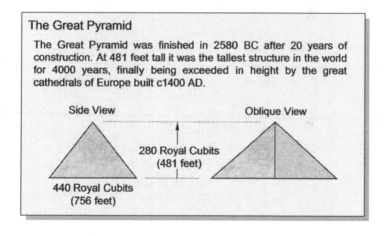

The Great Pyramid

The Great Pyramid was finished in 2580 BC after 20 years of construction. At 481 feet tall it was the tallest structure in the world for 4000 years, finally being exceeded in height by the great cathedrals of Europe built c1400 AD.

Side View

Oblique View

280 Royal Cubits
(481 feet)

440 Royal Cubits
(756 feet)

The Pyramids were constructed almost entirely of limestone blocks from local mines on the West bank of the Nile but some interior stones, for example stones over the pharaoh's tomb, are granite from Aswan, four hundred miles to the south. The Great Pyramid was covered in casing stones of white limestone mined at the Tura quarry on the East side of the Nile. The casing stones were beveled on the outer edge so that the pyramid sides were smooth and flat, not stepped as seen today.

Sir William Matthew Flinders Petrie

During 1880 through 1882 Sir[1] William Matthew Flinders Petrie (1853-1942 AD) conducted the definitive survey of the Great Pyramid. He measured the length of the four sides and the height of each of the 203 visible courses of stone to a high degree of accuracy.

Sir Petrie in 1903

Sir Petrie published his findings in *The Pyramids and Temples of Gizeh*, 1883. This book was republished online

---

[1] Flinders Petrie was knighted in 1923.

in 2003 at *The Pyramids and Temples of Gizeh Online* by Ronald Birdsall with a revision in August 5, 2012; the online address is <http://www.ronaldbirdsall.com/gizeh>

Casing Stones

The white Tura limestone casing stones survived 2100 years of Egyptian rule, 1200 years of Persian, Greek, and Roman rule, and another 700 years of Arab rule. So why were they removed? We can blame it on an insult and the bubonic plague as follows.

In 1206 AD Genghis Khan (1162-1227 AD) formed the Mongol Empire. Genghis tried to establish trade routes along the "Silk Road" with Iran (Iran was referred to as Khwarezmia at that time). The Silk Road was established c200 BC when the Chinese traded silk with the European countries. When Khan's ambassadors arrived at Iran they were more than rudely treated by the Shah, some were shaved and some beheaded. Since ambassadors were supposed to be inviolable, this grievous insult resulted, in 1220 AD, in Genghis Khan assembling 200,000 soldiers and marching thousands of miles to conquer Iran and raze the capital city of Samarkand. Ultimately, Khan's grandson became leader of Iran and he set his sights on Babylonia on the western side of the Zagros Mountains (the Zagros Mountains form the border between Iran and Babylonia).

In 1258 AD Baghdad was sacked by the grandson of Genghis Kahn with the result that the center of Islamic culture moved from Baghdad to Cairo and Cairo then became a growing city.

In 1348 AD the Black Death (the bubonic plague) arrived in Egypt (it originated in the plains of central Asia and travelled west along the Silk Road) killing 40% of the population, not only in Egypt but throughout Europe. As a result of the plague, there were insufficient workers to mine the quarries at a rate needed to build fortifications against the threat of Mongol expansion. However, there were 21 acres of pure white Tura limestone 3 to 6 feet thick on the great pyramid. Therefore, in 1356 AD, the casing of the Great Pyramid was removed by the Bahri Sultan An-Nasir Nasir-ad-Din al-Hasan for building projects in Cairo. Some of these stones can still be seen in Cairo.

Great Pyramid Design

Aesthetic Considerations

An aesthetic consideration that the architect of the Great Pyramid may have been concerned with is the visual tradeoff between the side view and the oblique view. Shown here are three possible designs with different rise to run ratios.

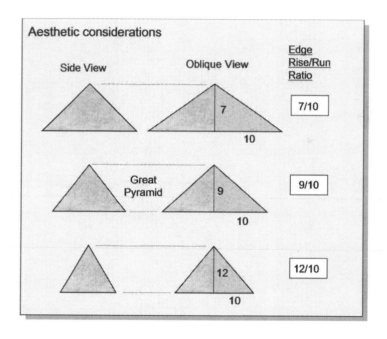

In the top diagram, the oblique view appears too squat and also the volume is 65% greater than the design in the middle (it would take 10 years longer to build). In the bottom diagram, the side view appears too peaked. The actual Great Pyramid is the middle design which has an oblique view rise to run ratio of 9/10. And given a side of 440 cubits, the resulting height will be 280 cubits.

Pyramid Shape, Equivalent Methods

The shape of the Great Pyramid can be arrived by any of the following three methods. These methods produce pyramids that are similar within a part in a thousand and would be essentially the same from an ancient Egyptian practical point of view. The side angles are the same to within 1/60 degree and the height of the Great Pyramid would be the same within 3 inches.

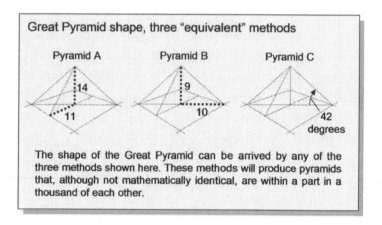

Great Pyramid shape, three "equivalent" methods

Pyramid A   Pyramid B   Pyramid C

The shape of the Great Pyramid can be arrived by any of the three methods shown here. These methods will produce pyramids that, although not mathematically identical, are within a part in a thousand of each other.

Pyramid A is constructed based on a side slope ratio of 14/11.

Pyramid B is constructed based on an edge slope ratio of 9/10.

Pyramid C is constructed based on an edge angle of 42 degrees.

Which method was employed is a matter of conjecture.

As an example of how similar these approaches are, if you built a model pyramid 10 feet high, the pyramids based on

Approach A and Approach B would differ in height by 6/1000 inch, about the thickness of a piece of paper; whereas Approach C would differ in height by about 1/16 inch.

A Case for the Edge Slope Method

Building the Great Pyramid based on maintaining a specific edge angle makes sense. After building a square foundation, the builder is faced with placing the stones for the second course.

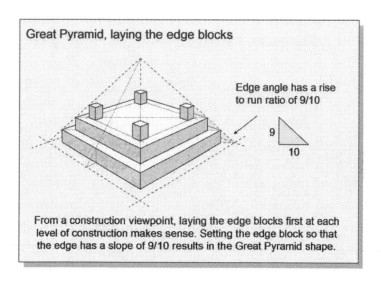

Great Pyramid, laying the edge blocks

Edge angle has a rise to run ratio of 9/10

9
10

From a construction viewpoint, laying the edge blocks first at each level of construction makes sense. Setting the edge block so that the edge has a slope of 9/10 results in the Great Pyramid shape.

As shown in the figure, a reasonable approach would be to place the corner stones on the diagonals of the square and then to set the stones into position such that the height of the stone relative to the distance from the corner is a particular ratio, in this example 9/10. A line would then be drawn from corner to corner as a guide in placing the remaining stones. The existence of edge sockets recessed into the bedrock, anchoring the edge casing, points to the special importance of the edges.

Assuming this was in fact the intention of the architect, a 9/10 ratio would result in a side angle of 51.84 degrees which is consistent with the measurements by Sir Petrie. In addition, assuming a 9/10 ratio and also assuming the architect wanted the base to be a square of exactly 440 cubits on a side, the height of the pyramid would be precisely (within 1/3 of an inch) 280 cubits which is also what Sir Petrie estimated.

A case for the side slope method

Detailed measurements of the Great Pyramid show that the core blocks along the baseline do not make a straight line from corner to corner but over the 756 foot length they deviate from a straight line by about 3 feet at the center and proportionately less as you move toward the corners. Therefore the casing in the middle of a side is 3 feet thicker ("thicker" meaning as measured horizontally form the outside toward the center) than at the edge of the pyramid. This suggests that the side center casing was installed as a guide, likely along with the edge casing, for positioning the rest of the casing. If you look carefully at the satellite view of the pyramid shown here you can see an indentation down the middle of the sides where presumably the casing was thicker.

If this approach was followed then the side slope of 14 rise to 11 run would have been used for the side casing and the edge slope of 9 rise to 10 run could have been used for the edge angle since they produce equivalent results.

Great Pyramid, side center casing

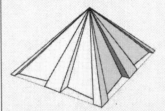

Detailed measurements of the Great Pyramid show that the core blocks along the baseline do not make a straight line from corner to corner but over the 756 foot length they deviate from a straight line by about 3 feet at the center and proportionately less as you move toward the corners.

Therefore the casing in the middle of a side is 3 feet thicker ("thicker" meaning as measured horizontally from the outside toward the center) than at the edge. The indentation of the baseline is greatly exaggerated in the illustration.

A Case for the 42 degree edge angle

The "palm" (four fingers) can be used as a convenient angular measurement. With your arm outstretched there are 15 palms in 90 degrees and therefore 6 degrees per palm and 1.5 degrees per finger. The ancients knew that if you stacked suns from the horizon to a point straight up (the zenith) it will take 180 suns; therefore, since there are 90 degrees from the horizon to the zenith, this amounts to ½ degree per sun diameter.

Also, as an aside, it takes the sun six hours (1/4 of a day) to travel 90 degrees. Since six hours is 360 minutes, that's 2 minutes per each of the 180 sun diameters. Since there are six degrees per palm it takes 24 minutes for the sun to travel through a palm angle. If, for example, the sun is then three palms (18 degrees) above the horizon, it will take one hour and twelve minutes to set.

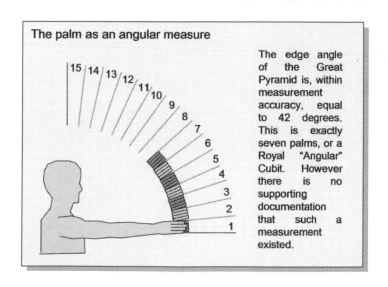

The palm as an angular measure

The edge angle of the Great Pyramid is, within measurement accuracy, equal to 42 degrees. This is exactly seven palms, or a Royal "Angular" Cubit. However there is no supporting documentation that such a measurement existed.

Seven palms are precisely 42 degrees which would be consistent with a Royal "Angular" Cubit if such a measurement technique existed. However, there is no supporting documentation for a Royal "Angular" Cubit. A 42 degree edge angle has a rise to run of 9.004/10 versus the 9/10 mentioned in the previous paragraph. If in fact such an angular measurement approach existed then the angle would be converted into a rise/run value for construction purposes and no doubt 9.004/10 would become 9/10. Regardless, it is interesting to envision the chief architect, when trying to get the Pharaoh to sign-off on the plan, pointing out that edge angle is precisely a Royal "Angular" Cubit.

Lastly, the angle from your eyes to a rainbow and back to the sun makes an angle of 42 degrees. This is an interesting coincidence. Other coincidences concerning the Great Pyramid's shape have to do with the Golden Ratio, Pi, and Orion's belt; these will be addressed in a later section.

Measurements

As mentioned earlier, the definitive survey of the Great Pyramid was conducted by Sir William Mathews Flinders Petrie (1853-1942 AD) and published in his book "The Pyramids and Temples of Gizeh" in 1886. His measurements of the four sides of the base, as measured along the seam between the pavement and the casing, averaged 9068.8 inches. Assuming that the intent of the design was a base of 440 cubits by 440 cubits then the average measured base length of 9068.8 inches results in 20.6109 inches per cubit.

Base Measurements of the Great Pyramid

Side	Inches	Resulting cubits per inch assuming 440 cubits was the intent
North	9069.4	20.6123
South	9069.5	20.6125
East	9067.7	20.6084
West	9068.6	20.6105
Avg.	9068.8	20.6109

Sir Petrie's slope measurements were more accurate on the North face and the weighted average of those measurements was 51 degrees 50 minutes and 40 seconds which is 51.8444 degrees. A slope very close to this value results from any of three different design approaches as shown in the table shown below. Approach A assumes a side slope with a rise to run ratio of 14/11. Approach B assumes an edge slope with a rise to run ratio of 9/10, and Approach C assumes an edge angle of 42 degrees. This edge angle of 42 degrees is exceedingly close to the angle produced by an edge ratio of 9/10.

You can see in the table that the height of the Great Pyramid is almost precisely 280 cubits regardless of the design approach. The three methods differ by no more than 3 inches out of 481 feet.

Nominal values of various Great Pyramid parameters are shown here. The base of the pyramid occupies 13.1 acres and the area of each side is 5.3 acres.

Great Pyramid Parameters

The Great Pyramid, parameters based on three different slope assumptions				
Asssuming 20.6109 inches per cubit and a square base of 440 cubits per side				
		A	B	C
Parameter	Units	14/11 side slope	9/10 edge slope	42 degree edge slope
edge slope rise/run		0.89995	0.90000	0.90040
side slope rise/run		1.27273	1.27279	1.27336
base (b)	cubits	440.000	440.000	440.000
base (b)	feet	755.733	755.733	755.733
base (b)	inches	9068.80	9068.80	9068.80
half diagonal (d/2)	feet	534.384	534.384	534.384
height (h)	cubits	280.000	280.014	280.140
height (h)	feet	480.921	480.946	481.161
height (h)	inches	5771.05	5771.35	5773.94
side angle (sa)	deg	51.84277	51.84419	51.85669
side length (sl)	feet	611.611	611.630	611.800
edge angle (ea)	deg	41.98576	41.98721	42.00000
edge length (el)	feet	718.9236	718.9400	719.0845

Great Pyramid Nominal Values

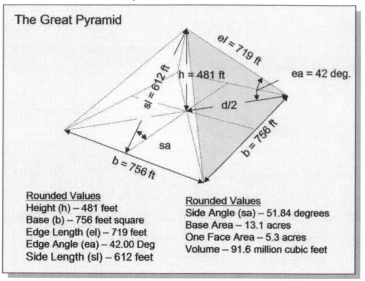

The Great Pyramid

el = 719 ft

sl = 612 ft   h = 481 ft

ea = 42 deg.

d/2

sa

b = 756 ft

b = 756 ft

<u>Rounded Values</u>
Height (h) – 481 feet
Base (b) – 756 feet square
Edge Length (el) – 719 feet
Edge Angle (ea) – 42.00 Deg
Side Length (sl) – 612 feet

<u>Rounded Values</u>
Side Angle (sa) – 51.84 degrees
Base Area – 13.1 acres
One Face Area – 5.3 acres
Volume – 91.6 million cubic feet

Courses

The survey by Sir Petrie measured the elevation (i.e. height) at the top of each course on both the northeast edge of the pyramid and on the southwest edge of the pyramid; this was up to the 203rd course on the northeast edge and up to the 201st course on the southwest edge. The courses were in general all different thicknesses with those at the bottom being about twice as thick as those at the top. As you can see in the illustration, the top of the pyramid was missing many courses at the time of the survey as it is today. This amounts to 15 courses if the missing courses were approximately the same thickness as the nearby preceding courses, but there was likely a cap stone, maybe made of granite, of perhaps 4 courses thick.

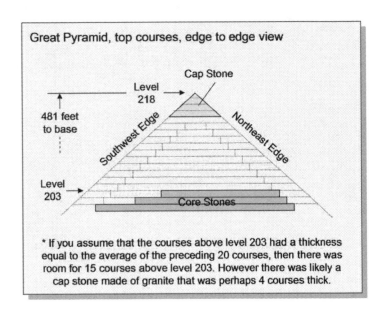

Great Pyramid, top courses, edge to edge view

* If you assume that the courses above level 203 had a thickness equal to the average of the preceding 20 courses, then there was room for 15 courses above level 203. However there was likely a cap stone made of granite that was perhaps 4 courses thick.

Construction

Building Blocks

The Great Pyramid was built on top of leveled bed rock
that was topped with an approximately 2 foot thick
limestone pavement that extended beyond the pyramid
30-40 feet; the exact distance is estimated from the
amount of rock-cut bed since the pavement has been
removed around the periphery. The core stones of the
pyramid are limestone from open pit mines on the west
bank of the Nile and are of a poorer quality than the casing
of pure white Tura limestone mined from tunnels on the
east side of the Nile. To a large extent the casing was
removed in 1356 AD but enough remained for Sir Petrie to
measure the base of the pyramid along the length where
the casing meets the pavement. In the four corners the
casing is set into "sockets" in the rock-cut bed.

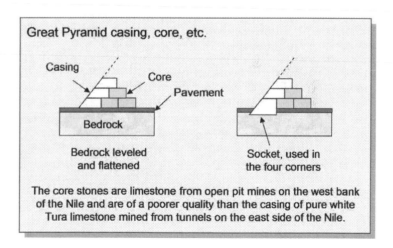

Great Pyramid casing, core, etc.

Casing

Core

Pavement

Bedrock

Bedrock leveled
and flattened

Socket, used in
the four corners

The core stones are limestone from open pit mines on the west bank
of the Nile and are of a poorer quality than the casing of pure white
Tura limestone mined from tunnels on the east side of the Nile.

Aligning with East-West

The Egyptians aligned the Great Pyramid relative to North with an accuracy of 1/10 degree, and it has been commented that they didn't even have a compass.

First, it is easy to align to North with a high degree of accuracy, and second you would never use a compass even if you had one since magnetic North is several degrees away from true North, and its location drifts with time.

Aligning the Great Pyramid with East-West

As shown here, you can establish an East-West line by marking the rising and setting moon (or sun)*. It shouldn't be difficult to position a pole right in the center of the rising moon within an accuracy of 1/5 of its diameter. Since the moon has an angular extent of ½ degree, 30 minutes, then 1/5 of this is 6 minutes of arc.

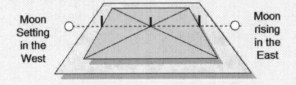

Moon Setting in the West

Moon rising in the East

* Once you have an East-West line it is easy to establish a North-South line perpendicular to it.

Number of Blocks in the Great Pyramid

It is commonly noted in the literature that Sir Petrie estimated that there are approximately 2,300,000 blocks in the Great Pyramid; the assumptions underlying this estimate are not known. However, we can come up with the same estimate by applying reasonable assumptions as shown here.

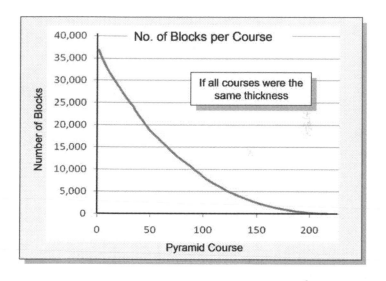

There are 203 courses visible on the Great Pyramid; 27 feet, about 15 courses, of structure was removed. If all courses were the same thickness then the number of blocks per course would decrease smoothly as the pyramid tapered toward the top as shown here. However, the course thickness is far from constant as shown in the figure below. The thickness depended to some extent on the natural stratification of the limestone quarry. For example, the bottom course is about 5 feet thick; this reduces, over the next 20 courses to 2 feet which then increases to 4 feet before reducing again. The actual per course variation results in an irregular blocks-per-course.

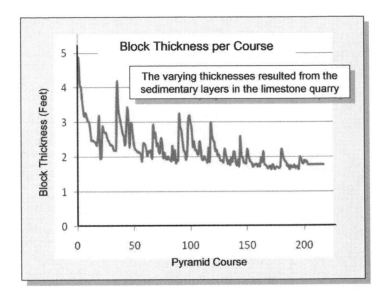

If you assume the blocks are the same shape but vary in size according to the course thickness you can estimate the number of blocks in each course, and the number estimated depends strongly on the assumed shape of the block.

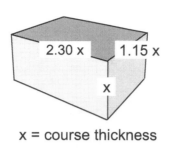

2.30 x   1.15 x

x

x = course thickness

The block shape shown here results in the same 2.3 million blocks estimated by Sir Petrie. Other shapes could of course result in the same total count as long as the volume per block does not change. Since the volume of all the pyramid courses is approximately 91.6 million cubic feet and since there are 2.3 million blocks, the average block is about 40 cubic feet which results in an average thickness of about 2.5 feet which is consistent with the earlier diagram. The average block, at 40 cubic feet, would weigh 6000 pounds at 150 lbs/cf for limestone.

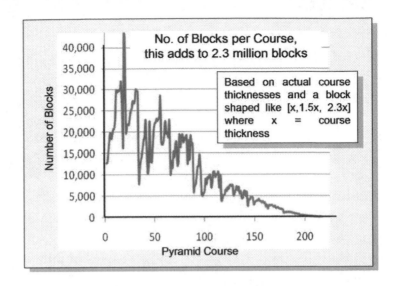

No. of Blocks per Course, this adds to 2.3 million blocks

Based on actual course thicknesses and a block shaped like [x,1.5x, 2.3x] where x = course thickness

Construction Time

Herodotus, "the Father of History", who published "The Histories" c420 BC, recorded the following after his visit to Egypt where he learned Egyptian history from the scribes:

1) It took 20 years to complete the Great Pyramid;
2) Levers were used to lift the blocks from one level to the next and so on up the pyramid; and
3) The upper levels were finished first.

This information was already 2000 years old at time of Herodotus' visit so the possibility exists that the information was not accurate. Re: Item 3) which states that the upper levels were finished first, I take to mean that the smooth finishing stones were set in starting from the top; which makes sense as this way you would have the stepped blocks to use a base as moved down the pyramid.

Using the 20 year figure, the pyramid construction project can be divided into the following phases:

Design	2 years
Site Preparation	2 years
Core Blocks	12 years
Casing	4 years

Assuming 16 work crews on each side (which allows almost 50 feet along the side for each crew at the base) for a total of 64 crews working at the same time, and assuming that each crew can position a block every hour, this would result in 64 blocks per hour, or 512 per 8 hour day. At this rate the 2,300,000 blocks would take 12.3 years. Of course you can vary the assumptions to result in different answers, e.g.:, assuming more crews, 12 hour days, time off during harvesting, etc.

Use of Ramps and Levers

Herodotus reported that the Egyptians stated that levers were used to raise the blocks. Despite this report, many people have suggested that ramps were used. Ramps seem like a reasonable approach for the lower blocks but ramps become prohibitively expensive as the pyramid gets higher. The figure shows levers being used to raise blocks, this is quite practical, fast, and well within the technical capability at the time. With a 15 to 1 advantage lever arm (5 feet long with a the lift end 4 inches from the pivot) a 100 pound force translates to a 1500 pound lift force, with four levers, one on each corner, the total lifting force is 6000 pounds. Each lift raises the block 3 inches, after which the block is secured at that height and the lever mechanism is raised 3 inches. It is easy to envision a mechanism with numerous pivot points and block holding points so that the block can be quickly lifted.

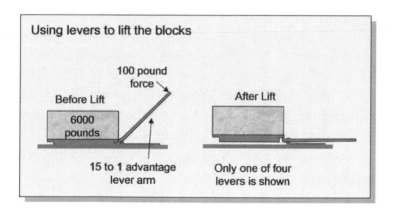

Using levers to lift the blocks

100 pound force

Before Lift

6000 pounds

15 to 1 advantage lever arm

After Lift

Only one of four levers is shown

Speculations

Was the Great Pyramid Based on Pi?

Pi is the number of times a diameter wraps around its circle. An excellent estimate for Pi is 3 1/7[1] which is accurate to within a part in a thousand.

3 1/7 = 3.1428

Pi = 3.1416

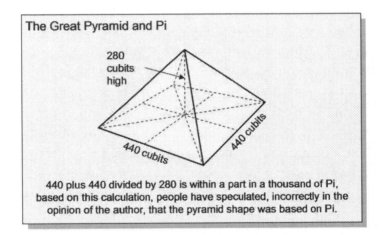

The Great Pyramid and Pi

280 cubits high

440 cubits

440 cubits

440 plus 440 divided by 280 is within a part in a thousand of Pi, based on this calculation, people have speculated, incorrectly in the opinion of the author, that the pyramid shape was based on Pi.

3 1/7 is the same as 22/7 so any shape that has numbers 22 and 7 or some multiple of those numbers can closely be related to Pi. Noticing that the pyramid is 440 cubits wide and 280 cubits high, it is not hard to come up with a relationship equal to 22/7, i.e., (440 + 440)/280 is the same as 22/7. Therefore, people have speculated that the pyramids are based on Pi.

---

[1] Pi is actually a never ending, never repeating decimal. The first 8 digits are 3.1415926 . . .

Although the Egyptians didn't use the formula A = Pi x r^2 to find the area of a circle, we can determine their equivalence to Pi from their procedure for calculating the area of a circle. The Egyptians estimated the area of a circle by taking 8/9 of the diameter (D) of the circle and squaring it to get the area (A) of the circle. I.e.: A = (D x 8/9)^2, or A = (2r x 8/9)^2, where r is the radius of the circle. This is equivalent to A = (256/81) x r^2, and, since the actual formula is Pi x r^2, the Egyptian equivalent for Pi is 256/81 (or 3 13/81) which is 3.1605 versus the actual value of 3.1416. The Egyptian value for Pi is too large by 0.6%.

The Fact that the Egyptians did not have an estimate for Pi equivalent to 22/7 suggests that they did not base the shape of the Great Pyramid on Pi, or on an estimate of Pi.

Was the Great Pyramid based on the Golden Rectangle?

There is much literature about the Golden Rectangle being the most pleasing shape for a rectangle. It is the only rectangle that maintains its shape when a square is added to its long side; when we study the Ancient Greeks we will learn how to construct this rectangle.

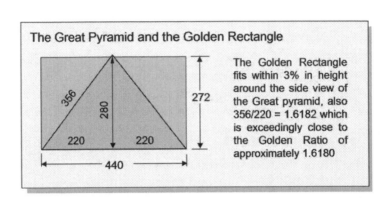

The Great Pyramid and the Golden Rectangle

The Golden Rectangle fits within 3% in height around the side view of the Great pyramid, also 356/220 = 1.6182 which is exceedingly close to the Golden Ratio of approximately 1.6180

There is also much speculation about whether or not the Egyptians may have based the shape of the Great Pyramid on the Golden Rectangle. In fact, there are no indications that the Egyptians knew anything about the Golden Rectangle, so why the speculation? It turns out that the Golden Rectangle fits around the side view of the Great pyramid quite nicely as shown in the figure, within 3%. Also, if you divide the pyramid side slant length of 356 cubits by half the base (356/220), this equals 1.6182 which is exceedingly close to the Golden ratio value of 1.6180.

However, these are just coincidences of math which fall out automatically if you have a pyramid with a side view rise/run of 14/11 or an oblique view rise/run of 9/10.

Were the Pyramids of Giza positioned like Orion's Belt?
The constellation Orion is shown here. People have observed that the spacing of the stars in Orion's belt is similar to the spacing of the three pyramids on the Giza plateau, and have suggested that this was a deliberate plan of the Ancient Egyptians.

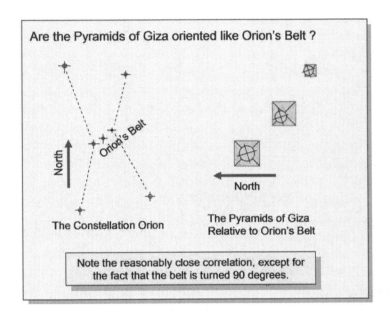

The figure also shows the three pyramids looking down from above and scaled in size and oriented as closely as possible to Orion's belt. There obviously is a close fit. However, note the direction of North in both figures, the pyramids are 90 degrees out of orientation relative to North. Rather than being based on Orion's belt, it is more likely that the first pyramid was located in a convenient spot and the other pyramids were placed an appropriate distance away and in a direction consistent with the local geography.

Babylonia

Geography
In this book the words Babylonia and Mesopotamia are used synonymously. Mesopotamia consists of the lands between and adjacent to the Tigris and Euphrates rivers; mostly modern Iraq, Syria, and eastern Turkey. The word "Mesopotamia" is from the Greek "Meso" for between and Potomia for "rivers".

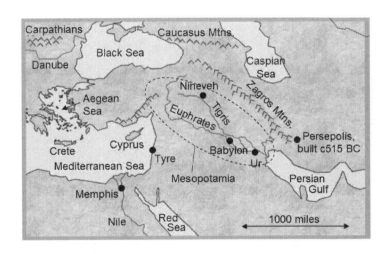

Peoples of Mesopotamia
In ancient times the southern lands near the Persian Gulf were populated by the Sumerians, the middle lands by the Babylonians and the northern lands by the Assyrians. Although at any given time, the whole region was under the control of one or another of these peoples and occasionally others. Being on the crossroads between Europe, Asia, and Africa, this region has experienced many

shifts of power and incursions of outside forces. The Sumerians lived in the Tigris-Euphrates river delta region. Based on linguistic analysis, the Sumerians have not been associated with any other peoples. c3300 BC writing systems were developed in Sumeria, c3000 BC standards of measurement and number systems were developed, c2700 BC the Royal Cubit was defined as 30 fingers (20.4 inches) in Sumeria and 7 palms of 4 fingers each (20.6 inches) in Egypt.

People in and around Mesopotamia

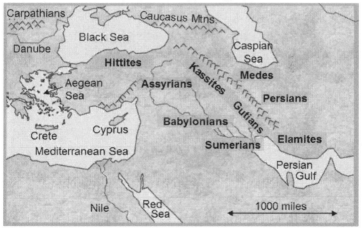

Equally ancient with the Sumerians was the nearby country of Elam with written records dating to 2700 BC. Elam was located along the Karun River, which flows out of the Zagros Mountains and then east along the Persian Gulf. Interestingly enough, the Sumerians and the Elamites, despite their close proximity, do not appear to be related since their languages do not seem to share a common origin. Several hundred miles upriver from the

Sumerians were the Babylonians; and several hundred miles upriver from the Babylonians were the Assyrians. The Babylonians and Assyrians, Semitic nations, became a power in the region a thousand years after the Sumerians.

Following the Babylonians and the Assyrians c2000 BC came Indo-European speakers who migrated south over the Caucasus and southeast of the Caspian Sea. These people included the Hittites, the Medes, and the Persians.

Also shown in the figure are the Gutians and the Kassites who lived in or along the Zagros Mountains, they also play an important role in Mesopotamian history. These people have not been linked to any other people.

Ancient Babylonian Timeline

Historical Periods

Shown here is an overview of the 5000 years of Mesopotamian history divided into three periods, Pre-Dynastic Period, Early Dynastic Period, and Empire Period.

Mesopotamian History Periods			
5300 BC	2900 BC	2270 BC	330 BC
Pre-dynastic Period Ubaid Uruk Jemdet Nasr 5300- 4000- 3100- 4000 3100 2900	**Early Dynastic Period**	**Empire Period**	
Increasing centralization of the population from villages to towns to cities. 5300 BC early settlements in Sumeria	Walled cities with kings and a social structure.	Central government that ruled over lands that extended 1000 miles from the Persian Gulf to the Mediterranean.	
	Bronze Age (3300-1200 BC)	Iron Age (1200 - 500 BC)	

The Pre-Dynastic Period lasted for 2400 years during which population centers increased in size from villages to towns to cities. This was followed by the 600 year Early Dynastic Period during which kingdoms and walled cities with all of the necessary social structure evolved. Following the Early Dynastic Period was the Empire Period which lasted 1940 years. During the Empire Period all of Mesopotamia came under the rule of one or another centralized government. The location of the capital and the peoples in control of the government changed many times (the longest government lasted about 400 years).

The people who formed the first empire were the Akkadians and they gave the name Sumer to the delta region.

The Empire Period

A timeline of Mesopotamian Empire Period is shown here divided into nine smaller periods[1]. Although there are nine periods, there are only six "empires", namely: Akkadian Empire, Ur III Empire, Babylonian Empire, Neo-Assyrian Empire, Neo-Babylonian Empire, and the Persian Empire. These empires all ruled lands extending from the Persian Gulf to the Mediterranean.

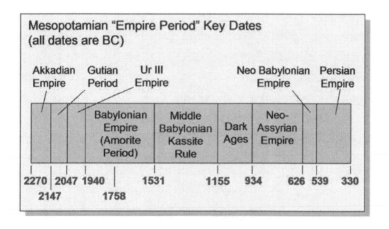

Akkadian Empire, 2270 BC

The Akkadians were a Semitic speaking people who lived in central Mesopotamia with their main city being Akkad (near Babylon) located 200 miles upriver from the Sumerians. Sargon "the great" of Akkad conquered Sumer in 2270 BC combining central and southern Mesopotamia

---

[1] Dates are based on the "short chronology" which assumes the sack of Babylon by the Hittites occurred in 1531 BC.

forming the Akkadian Empire. Subsequent conquests resulted in all of Mesopotamia being included in his empire. According to legend, Sargon was conceived in secret and put in a reed basket in the river; a very similar story to that of Moses. The Akkadian Empire lasted 123 years before it was weakened due a severe drought (Bond[1] Event 3) and Akkad was conquered by the Gutians. The Akkadian language became the lingua franca of Mesopotamia for over 1000 years and it was still being used by scribes for another 1000 years.

The Gutians, 2147 BC

The Gutians, from the central Zagros region, destroyed Akkad in 2147 BC bringing an end to the Akkadian Empire and initiating the "Gutian Period". During this period, the Amorites from Syria started migrating into Mesopotamia and the Hittites migrated over the Caucasus and settled in northern Anatolia. The Gutians were defeated in battle by the rulers of Ur, bringing in the Ur III Empire.

The Ur III Empire, 2047 BC

The Ur III Empire began in 2047 BC when the rulers of Ur defeated the Gutians. The Ur III Empire initiated the Sumerian Renaissance. The Ur III Empire eventually extended from the Persian Gulf to the Mediterranean Sea. The northern regions of the empire, near modern Syria, were occupied by a nomadic people known as the

---

[1] Bond Events are climate fluctuations that have been observed to occur approximately every 1500 years since the end of the last ice age. They were proposed in 1997 by Gerald C. Bond (1940-2005 AD); he theorized that the events were caused by a 1500 year solar cycle.

Amorites. Following an Amorite revolt c1970 BC a 170 mile wall was constructed between the Tigris and Euphrates to hold back the Amorites. The rulers of Ur were defeated by the Elamites allowing the Amorites to gain power. C2000 BC the Epic of Gilgamesh was written and also about that time the Great Ziggurat of Ur was built. Ziggurats were tall stepped pyramid shaped buildings with square or rectangular bases used for religious purposes at many locations in Mesopotamia. Since stone was not readily available in the delta region of the Tigris-Euphrates, Ziggurats were constructed of kilned mud bricks using hot bitumen (tar) as mortar. Bitumen was readily available as it seeped from the ground in numerous areas in the Tigris-Euphrates valley. The Great Ziggurat of Ur had a base of about 200 feet on its longest side. Another famous ziggurat was the ziggurat of Babylon built c1800 BC. According to the Greek historian Herodotus, writing in 440 BC, the Babylon ziggurat was: "a tower of solid masonry, a furlong [1/8 mile or 660 feet] in length and breadth, upon which was raised a second tower, and on that a third, and so on up to eight. The ascent to the top is on the outside, by a path which winds round all the towers. When one is about half-way up, one finds a resting-place and seats, where persons are wont to sit some time on their way to the summit. On the topmost tower there is a spacious temple, and inside the temple stands a couch of unusual size, richly adorned, with a golden table by its side." This quote is from the 1860 translation by George Rawlinson. The Babylonian ziggurat is believed by some to be the "Tower of Babel" mentioned in the Bible.

The Amorite Period, 1940 BC

When the Elamites defeated the rulers of Ur in 1940 BC, the Amorites, a Semitic speaking people from the area of modern Syria and the Levant, who had migrated south from Syria into Mesopotamia, took control, starting 400 years of Amorite rule. In 1728 BC Hammurabi became the first king of the Babylonian Empire by conquering neighboring regions including Assyria and the lands near the Zagros Mountains. He remained king for 42 years. The Code of Hammurabi (a list of 282 laws) was written c1758. The Babylonian creation myth, the Enuma Elish, is dated to this period. The Enuma Elish is a story of creation very similar to that found in Genesis. Babylon was sacked by the Hittites in 1531 BC (after which the Hittites returned home) and the Kassites from across the Tigris near the Zagros Mountains took control.

Kassite Rule, 1531 BC

The Kassites from the Zagros Mountain region gradually took control after Babylon was sacked by the Hittites in 1531 BC. The Kassite Period lasted according to tradition for 576 years, i.e. well into the Mesopotamian Dark Ages, but as shown here it lasted almost 400 years. The Kassites introduced horses into the region. The date of the sack of Babylon is a key milestone to establishing the earlier chronology of Mesopotamia. Various chronologies are used by scholars depending on their views of supporting evidence. The dates used herein are the "short chronology"; the opinions of scholars as to the date of the sack of Babylon can vary by over 100 years.

Mesopotamian Dark Ages, 1155 BC

The year 1200 BC marks the end of the Bronze Age and the beginning of the Iron Age. Approximately at this time there occurred what is referred to as the Bronze Age Collapse which included the destruction of many cities along the eastern Mediterranean and the migration of many people. Also, there were invasions by the "Sea People" (the origin of the Sea People is not known). 1200 BC marks the end of Mycenaean Greece and the start of the Greek Dark Ages. It also marks the collapse of the Hittite Empire. For the timeline shown here, the Mesopotamian Dark Ages began in 1155 BC when the Kassites were overthrown by the combined forces of Assyria and Elam. During the "Dark Ages" cities went into decline and records from this period are minimal.

The Neo-Assyrian Empire, 934 BC

In 934 BC Ashur-dan II of Assyria started the Neo-Assyrian Empire which maintained hegemony over the region for 300 years. In 729 BC Babylon was conquered by Assyria. In 720 BC Assyria conquered the Kingdom of Israel and the people were taken away resulting in the "Ten Lost Tribes of Israel". In 705 BC Sargon II died in a battle with the Cimmerians (an Indo-European people that lived near the Caucasus Mountains) and his son Sennacherib became king. Sennacherib moved the capital to Nineveh. In 701 BC the Kingdom of Judah revolted against Assyria and Sennacherib laid siege to Jerusalem but was forced to leave when his army became sick. In 689 BC Assyria destroyed Babylon by flooding the city. In 671 BC Assyria conquered Egypt. In 626 BC Babylon revolted bringing an end to the Neo-Assyrian Empire.

The Neo-Babylonian Empire, 626 BC

In 626 BC Babylon rebelled against Assyria, initiating actions that would form the Neo-Babylonian Empire. In 612 BC Babylon, in alliance with the Medes, sacked Nineveh and the seat of power in Mesopotamia returned to Babylon. In 604 BC Nebuchadnezzar becomes king of the Chaldeans. The Chaldeans, a Semitic people, had entered Sumeria 100 years earlier and soon become leaders. The Neo-Babylonian Empire is also referred to as the Chaldean Empire. In 597 BC Nebuchadnezzar captured Phoenicia and Judah and burned Jerusalem. In 587 BC the Jews from Judah were exiled to Babylon for 59 years (The Babylonian Exile). Nebuchadnezzar built the Hanging Gardens of Babylon, c600 BC, one of the seven wonders of the ancient world.

Hanging Gardens of Babylon, c600 BC

The Persian Empire, 539 BC

The Persian Empire was formed in 539 BC when Cyrus the Great of Persia captured Babylon. He released the Jews in 538 BC from their 59 year exile (597-538 BC) in Babylon and let them return home and rebuild their temple.

Shown here are the first four kings of Persia.

The first four kings of Persia	
Cyrus the Great (ruled 550-530 BC)	*Conquered Babylon in 539 BC, freed the Jews in 538 BC*
Son	
Cambyses II (ruled 529-522 BC)	*Conquered Egypt in 525 BC*
Darius I the Great (ruled 521-486 BC)	*Defeated by the Greeks in 490 BC in the battle of Marathon*
Son	
Xerxes I the Great (ruled 485-465 BC)	*Defeated by the Greeks in 480 BC in the battle of Salamis*

Cyrus the Great is considered one of the greatest rulers of all time. His policies benefited not only the empire but also the regions where he ruled. He was very tolerant of local customs.

Darius the Great was the third king of Persia. He built the capital at Persepolis c515 BC. In response to a revolt of Greek cities in Ionia (along the western coast of modern Turkey), he attacked Greece but was defeated in the battle of Marathon. His son Xerxes I decided to accomplish what his father could not using a two-pronged attack. Xerxes entered the lands north of Greece with a large army which was held at bay for several days by a small Greek force led by Sparta; this is known as the Battle of Thermopylae. The Persian land force then moved on to Athens which had been evacuated. The other prong of the Persian advance was a huge fleet which was defeated by the Greeks at the Battle of Salamis. In 330 BC the Persian Empire fell to Alexander the Great.

Babylonian Astronomy

Astronomical Concepts

The Celestial Sphere

The ancient Babylonian astronomers believed that the Earth was stationary flat disc and that the stars were all at the same distance from us attached to a "celestial sphere" which rotated completely around every 24 hours.

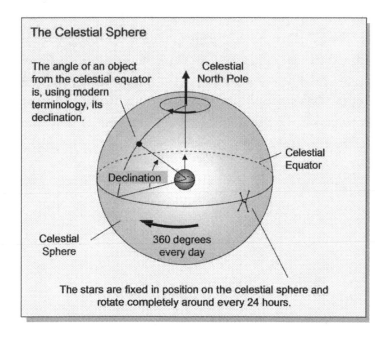

The Celestial Sphere

The angle of an object from the celestial equator is, using modern terminology, its declination.

Celestial North Pole

Declination

Celestial Equator

Celestial Sphere

360 degrees every day

The stars are fixed in position on the celestial sphere and rotate completely around every 24 hours.

In reality the stars are many different distances from us but from our point of view they are so far away that they appear fixed to the sphere. If you look in the direction of the North Star, you can see that stars move in circles around the celestial North Pole. The celestial equator and the Earth's equator are in the same plane.

In the illustration Orion is shown reversed since we are looking from "outside" the celestial sphere. The position of stars can be located on the celestial sphere by noting the angle north or south of the celestial equator and the angle east or west of a reference point. In modern terminology these angles are the declination and right ascension. Right ascension is measured eastwards from the vernal equinox along the celestial equator. This measurement approach is analogous to the use of latitude and longitude used to locate points on Earth; with the difference that longitude is measured with respect to the prime meridian (longitude) which passes through the Royal Observatory in Greenwich England. Ancient Babylonians however measured the position of stars relative to the ecliptic plane which we will learn about next.

The Ecliptic

The ancient astronomers observed that the stars complete a 360 degree circle every day, but the sun only moves 359 degrees a day. I.e., the sun appears to move through the celestial sphere at a rate of one degree per day[1] in a direction opposite to the stars. They also noticed that the path of the sun relative to the celestial sphere is on a plane that is tipped at an angle of 23.5 degrees relative to the celestial equator. The path of the sun is named the ecliptic because on this path eclipses occur.

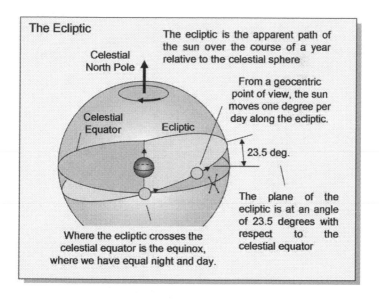

What actually is happening is the Earth is moving around the sun once per year giving the impression that the sun is

---

[1] One degree is equivalent to two sun diameters.

moving through the celestial sphere once per year. Once per year is very close to one degree per day.

Star Catalogs

The Babylonians compiled star catalogues starting c1200 BC. The catalogues listed the major stars and constellations along the ecliptic as well as near the North Pole. The Babylonians measured longitude in degrees counterclockwise from the vernal equinox and latitude in degrees above or below the ecliptic.

The Zodiac

The apparent path of the sun through the stars, as viewed from the Earth, is referred to, as mentioned above, as the ecliptic. Along this path are stars that remain fixed relative to each other and relative to the ecliptic. Shown here are the stars near the ecliptic. Since there are twelve full moons in a year, the Babylonians grouped the stars into 12 constellations resulting in the signs of the zodiac, as shown here. Note, Orion is shown for reference only, Orion is not a sign of the zodiac.

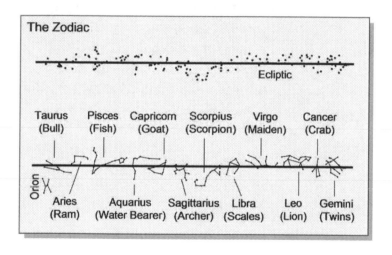

The Zodiac and the Sun Tell the Seasons

Knowing the location of the sun relative to the constellations establishes the time of year, which was used to predict cycles, such as when to plant in the spring, and when to celebrate religious events. Since you cannot see the stars when the sun is out, how does one know where the sun is relative to the stars? You can determine this by starting a timer when the sun is at the horizon and then waiting for two hours, the Earth will have turned through exactly thirty degrees, i.e. through one constellation. Then, having your map of the Zodiac, you can then locate the sun as being thirty degrees away from the point you noted after the two hour time interval. The Babylonians used water clocks which should have been able to time the two hour interval to within one minute, which equates to 1/4 degree accuracy in locating the sun.

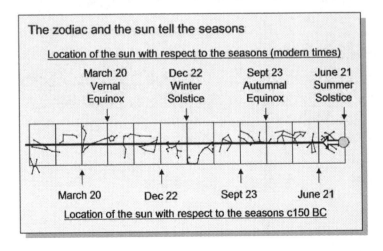

The Cause of the Seasons, a Geocentric View

In this section the cause of the seasons is explained from a Babylonian geocentric perspective, and in the next section the cause of the seasons is explained from a modern heliocentric perspective.

From a geocentric point of view, as believed by The Babylonians, the seasons are caused by the sun providing hotter and longer days in the summer and colder and shorter days in the winter, and these variations in the length of a day and the temperature are caused by the path of the sun moving north and south during the year as shown in the figure. They had no explanation as to why the path of the sun changed over the year; that would have to await development of the heliocentric view. In the middle of the summer the longest day occurs when the sun's path is at its furthest point north; this is the day of the summer solstice. The word solstice is from the Latin sol for Sun and sistere, to remain still. On the summer solstice the sun has stopped moving north and is not yet moving south, its path is standing still. The shortest day of the year is on the winter solstice. In between are the equinoxes (for equal night) when the day and night are of equal length. There are two equinoxes, one when the sun's path is travelling north, the Vernal equinox, and the other when the sun's path is travelling south, the Autumnal equinox. The time from the vernal equinox to the autumnal equinox is about 187 days whereas the time from the autumnal equinox to the vernal equinox is about

178 days. The Babylonians knew of this difference but they did not know the reason.

These equinox to equinox durations differ from exactly half a year by about 5 days. The cause of this would not be determined until the time of Kepler, c1600 AD, who calculated that the planetary orbits, including the Earth's, revolve about the sun in an Ellipse and that the orbital speed is slower farther away from the sun and faster nearer the sun. During the summer months the Earth is farther from the Sun than in the winter (but tilted toward the sun causing the Earth to be warmer in the summer). Being farther away in the summer, the orbital motion is slower during the summer months and therefore the time from vernal to autumnal equinoxes is longer than a half year. Since the Earth is farther away in the summer, it is closer to the Sun in the winter, which is counterintuitive, but this results in an autumnal to vernal period that is shorter than a half year.

It is noted that the Earth's axis of rotation precesses with a period of 26,000 years and there will come a time, thousands of years in the future, when the Earth is farther from the Sun in the winter causing more severe winters and helping to lead us into another ice age.

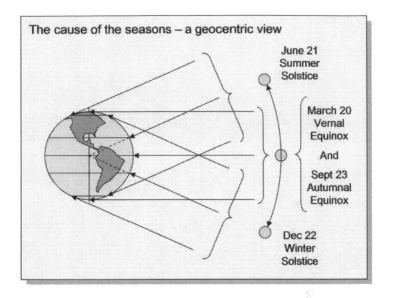

The cause of the seasons – a geocentric view

June 21
Summer
Solstice

March 20
Vernal
Equinox

And

Sept 23
Autumnal
Equinox

Dec 22
Winter
Solstice

The Cause of the Seasons, a Heliocentric View

The heliocentric view of planetary motion did not get established until after the renaissance with contributions from Copernicus, Galileo, Kepler, and Newton. What is really happening to cause the seasons is shown here. The Earth is orbiting around the Sun once per year but the Earth's axis of rotation is tipped at an angle of 23.5 degrees with respect to the orbital plane. So in summer the northern hemisphere is leaning towards the Sun bringing in the longer warmer days but in winter the northern hemisphere is leaning away from the Sun causing colder weather.

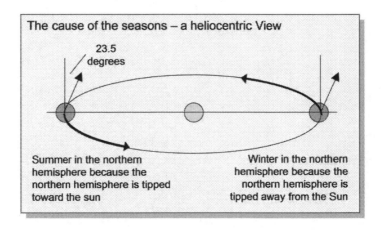

The cause of the seasons – a heliocentric View

23.5 degrees

Summer in the northern hemisphere because the northern hemisphere is tipped toward the sun

Winter in the northern hemisphere because the northern hemisphere is tipped away from the Sun

Origin of 360 Degrees

Dividing a circle into 360 degree was a Babylonian invention. Its origin is due to the fact that the sun moves through the celestial sphere at a rate of one degree per day, very close to 360 degrees per year. Also, full moon to full mood is about 30 days which divides the year nicely into twelve months.

It is interesting to note that the number 360 is easy to work with because it has a total of 24 divisors (1, 2, 3, 4, 5, 6, 8, 9, 10, 12, 15, 18, 20, 24, 30, 36, 40, 45, 60, 72, 90, 120, 180, and 360). Also, 360 is the smallest number that is divisible by every natural number from 1 to 10, except for 7.

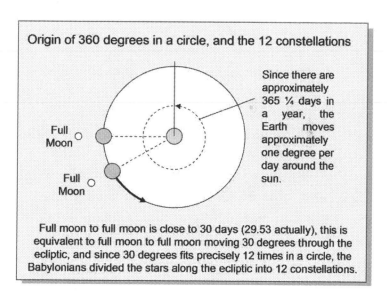

Origin of 360 degrees in a circle, and the 12 constellations

Full Moon

Full Moon

Since there are approximately 365 ¼ days in a year, the Earth moves approximately one degree per day around the sun.

Full moon to full moon is close to 30 days (29.53 actually), this is equivalent to full moon to full moon moving 30 degrees through the ecliptic, and since 30 degrees fits precisely 12 times in a circle, the Babylonians divided the stars along the ecliptic into 12 constellations.

Metonic Cycle

The Babylonians, c500BC discovered the Metonic cycle. This cycle is named after Meton, an Athenian astronomer, who flourished c430 BC and who proposed a calendar based on this cycle. The cycle is based on the fact that there are almost exactly 235 full moons in a 19 year period. If you have a full moon on a particular night, how close will you be to a full moon precisely a year later? The answer is shown cumulatively in the following chart. After one year there will have been 12 full moons with the last full moon being about 11 days prior to the end of the year. After two years the nearest full moon will be about 8 days after the end of the second year; and so on. Up to year 18, the closest alignment with the year end is within 1.5 days. However after 19 years the full moon is lined up with the year to within 2 hours. The basis for this is that the year has 365.24219 days and the lunar period is 29.530589 days. Hence, 235 times 29.530589 equals 6939.6884 days; and 19 times 365.24219 equals 6939.602 days; a difference of .086 days, about 2 hours. Knowing that 19 years is approximately 6940 days, the ancients could calculate the length of a year very accurately. The Hebrew calendar, which has a 19 year period, uses the Metonic cycle to regulate its intercalary months. Some of the characteristics of the Hebrew calendar are as follows.[1] "The Hebrew calendar has some months that are 29 days long and some that are 30 days long, so as to obtain an average of 29.5. Also, in order to ensure that Holidays do

[1] The paragraph in quotes is directly from the review comments provided by Dr. Dan Varon, thank you Dan.

not drift from the seasons in which they are supposed to be celebrated (e.g., Passover in the Spring), the Hebrew calendar is synchronized with the solar calendar by adding a 13th month to the year, 7 times in a 19-year cycle, in years 3,6,8,11,14,17, and 19."

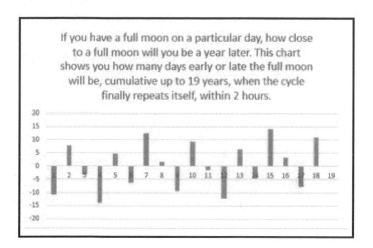

If you have a full moon on a particular day, how close to a full moon will you be a year later. This chart shows you how many days early or late the full moon will be, cumulative up to 19 years, when the cycle finally repeats itself, within 2 hours.

The Visible Planets

There are five planets in the night sky visible to the naked eye: Mercury, Venus, Mars, Jupiter, and Saturn. These were the only planets known to man until Uranus was discovered in 1781, Neptune in 1846, and Pluto in 1930. Pluto was reclassified in 2006 as a dwarf planet.

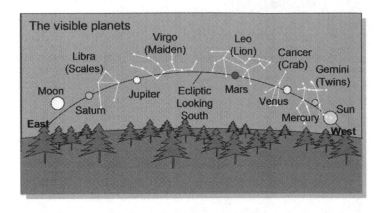

The visible planets

Virgo (Maiden) · Leo (Lion) · Libra (Scales) · Cancer (Crab) · Gemini (Twins) · Moon · Jupiter · Ecliptic Looking South · Mars · Venus · Saturn · Mercury · Sun · East · West

The planets, and the moon, all move on a path that is very close to the ecliptic. On a year to year basis, Mars, Jupiter and Saturn all move from west to east along the ecliptic. Saturn takes 29 years to orbit the sun and therefore moves approximately 12 degrees annually. Jupiter takes 12 years to orbit the sun and therefore year to year moves 30 degree in the night sky. Mars takes 1.9 years to complete its orbit and therefore moves 190 degrees annually. Mercury and Venus are closer to the sun than the earth. As viewed from the Earth, Mercury is always within 28 degrees of the sun and Venus is always within 48 degrees of the sun.

The arrangement of the planets shown in the figure is just for illustration purposes; more often than not, some planets are not visible because they are below the horizon. When the planets do come close to one another it is called a conjunction.

Summary

We have seen that the ancient Egyptians established fundamental mathematical procedures needed to conduct government business.   They had standard rules for addition, subtraction, multiplication, and division. They could determine the area of a parcel of land for tax purposes. They could calculate the volume of cylinders in which to store grain. They could determine the volume of a pyramid, including a truncated pyramid, so that they could calculate cost and estimate construction times. Their building projects required a well thought out design and required established stone mining and manufacturing abilities. The mathematical procedures were all cook book in nature and since they were sufficient, and since the Egyptian culture and religion discouraged change, the Egyptians did not try to generalize or advance their mathematical concepts; that would have to await the Greeks.

The Babylonians were the first to make detailed astronomical observations and record them in a star catalog. They identified the ecliptic, the path of the sun through the stars, and measured the location of stars relative to the ecliptic. They divided the circle into 360 degrees because the sun moves through the celestial sphere at a rate close to one degree per day, very close to 360 degrees per year. Also, full moon to full moon is about 30 days which divides the year nicely into twelve months. Therefore, they divided the sky along the ecliptic into twelve thirty-degree segments and grouped the nearby stars into twelve constellations comprising the signs of the zodiac.

Acknowledgements

Thank you to my wife Daria[1] for providing a loving and comfortable home in which to create this book. Thank you to my friends and family who have taken the time to read various portions of this series and who have always provided constructive encouragement. A special thanks to Dr. Dan Varon who carefully read this book and provided many helpful suggestions. The cover background is based on a public domain image of the Orion Nebula by NASA. The various pictures, not the illustrations, are from Wikipedia and are covered by the GNU Free Documentation License, thank you to those persons who provided the originals.

---

[1] Daria unfortunately passed away on June 22, 2013 at the age of 66 after a fight with lung cancer (she never smoked).

Appendices

Formatting

Text

Text for this book was generated using Microsoft Word with font set to Calibri 11 pt and with footnotes set to Calibri 10 pt. For conversion to Kindle format, the Word file was saved as *Web Page, Filtered (*.htm, *.html)* and then converted to ".mobi" using calibre, an application developed by Kovid Goyal, thank you Kovid.

Illustrations

Illustrations were generated in PowerPoint enclosed in a rectangle 5.4 inches wide, filled with a light color, with shadow right and bottom. The text within the illustrations uses Arial 12 pt font except for the title which is Arial 14 pt. Illustrations were converted to .jpg, 300 dpi, and inserted into Word *"In Line"* and *"Centered"*.

Paperback book formatting

If you are holding a paperback version of this book, it was prepared at createspace.com, an Amazon company. Paper size was set to an industry standard width of 5.06 inches and height of 7.81 inches. Margins were set as follows: .5 top, .6 bottom, .7 inside, and .7 outside resulting in a print area of 3.66 x 6.71. The .7 side margins allow adequate inside space without setting a "gutter". Text was set to

Calibri 11 pt with footnotes at Calibri 10 pt. Images were reduced, after inserting into the text, from the original width to match the text width of 3.66 inches. Cover artwork was created using PowerPoint, and then pasted into Word as an enhanced metafile, or as a jpg file, and then the Word document was saved as a .pdf file for upload to createspace.com.

Algebra Refresher

1. Algebra is shorthand for stating relationships. For example, instead of stating that the area of a rectangle equals the length of the base times the height, you can write an equation: a = bh, each letter represents a number. The equal sign "=", means that everything to the left of the equal sign is equal to everything to the right of it.

2. By convention, when two letters are written together, the numbers are intended to be multiplied. So "bh" means "b" times "h".

3. When the same letter is multiplied by itself, another shorthand is used. Instead of: y = aa, you write $y = a^2$ which is read as y equals a squared.

4. Once you have an equation, you can "operate" on it. As long as you perform the same operation on both sides of the equation, then the equation remains valid. For example, if a/b = b/c, then if you multiply both sides by b, the result is $a = b^2/c$.

5. Symbols used for various operations are:

  x or * for multiplication
  + for addition
  - for subtraction
  / for division
  √ for square root

6. Parentheses are used to group letters to make an equation simpler or clearer. For example:
$z = 1/d+e$, is clearer if you write,
$z = 1/(d+e)$, assuming that is what was intended.

It should be noted that there is a precedence of operations as follows: terms inside parentheses; exponents and roots; multiplication and division; and, addition and subtraction. According to this precedence an expression may be correct as displayed, but parentheses often help to ensure clarity.

7. Numbers that are multiplied together are referred to as factors. Given an equation with multiple terms and a common factor, e.g.: $y = ab+ac$, the common factor (in this case "a") can be "factored out" and the equation displayed as follows:
$y = a(b+c)$

8. Powers and Roots
Powers: $(Y \times Y) \times (Y \times Y \times Y) = Y^2 \times Y^3 = Y^5$; or, in general $Y^n * Y^m = Y^{(n+m)}$

Roots: $\sqrt{(a \times b)} = \sqrt{a} \times \sqrt{b}$

END